Trickle Bed Reactors

Trickle Bed Reactors
Reactor Engineering & Applications

Vivek V. Ranade, Raghunath V. Chaudhari,
Prashant R. Gunjal

Amsterdam • Boston • Heidelberg • London • New York • Oxford
Paris • San Diego • San Francisco • Sydney • Tokyo

ELSEVIER

Elsevier
The Boulevard, Langford Lane, Kidlington, Oxford, OX5 1GB, UK
Radarweg 29, PO Box 211, 1000 AE Amsterdam, The Netherlands

British Library Cataloguing in Publication Data
A catalogue record for this book is available from the British Library

Library of Congress Cataloging-in-Publication Data
A catalog record for this book is available from the Library of Congress

ISBN: 978-0-444-52738-7

For information on all **Elsevier** publications
visit our web site at elsevierdirect.com

Printed and bound in Spain

11 12 13 14 10 9 8 7 6 5 4 3 2 1

Working together to grow
libraries in developing countries
www.elsevier.com | www.bookaid.org | www.sabre.org

ELSEVIER   BOOK AID
International   Sabre Foundation

Contents

Preface   ix

1. **Introduction**   1
   Trickle Bed Reactors   1
       Basic Configurations and Operation of Trickle Beds   5
       Comparison with Other Reactors and Applications   9
   Reactor Engineering of Trickle Bed Reactors   12
       Key Issues   13
       Multiscale Approach for Reactor Engineering
       of Trickle Beds   16
   Organization of this Book   20
   References   22

2. **Hydrodynamics and Flow Regimes**   25
   Introduction   25
   Flow Regimes   26
       Trickle Flow Regime   29
       Pulse Flow Regime   30
       Spray Flow Regime   31
       Bubbling Flow Regime   31
   Flow Regime Transition   32
   Estimation of Key Hydrodynamic Parameters   40
       Pressure Drop   40
       Liquid Holdup   45
       Wetting of the Catalyst Particles   49
       Gas–Liquid Mass Transfer Coefficient   55
       Liquid–Solid Mass Transfer Coefficient   57
       Gas–Solid Mass Transfer   59
       Axial Dispersion   60
       Heat Transfer in Trickle Bed Reactors   63
   Summary   68
   References   69

3. **Reaction Engineering of Trickle Bed Reactors**   77
   Introduction   77
   Overall Rate of Reaction   79
       Completely Wetted Catalyst Particles   80
       Partially Wetted Catalyst Particles   86
       Exothermic Reactions   89

Reactor Performance Models for Trickle Bed Reactors   91
  Empirical Pseudo-homogeneous Models   102
  Generalized Model for Complete Wetting of Catalyst Particle 103
  Adiabatic Trickle Bed Reactor Model   104
  Non-isothermal Trickle Bed Reactor Model: Complex Reactions 105
  Periodic Operations in Trickle Bed Reactors   108
Summary   112
References   112

4. Flow Modeling of Trickle Beds   117
  Introduction   117
  Characterization of Packed Beds   119
    Randomly Packed Bed   120
    Structured Bed   124
  Single-Phase Flow through Packed Bed   126
    Modeling Approaches   128
    Model Equations and Boundary Conditions   130
    Flow through an Array of Particles   132
    Flow through a Packed Bed of Randomly Packed Particles   142
  Gas–Liquid Flow through Packed Beds   145
    Modeling of Gas–Liquid Flow through Packed Bed   146
    Simulation of Gas–Liquid Flow in Trickle Beds   154
    Simulation of Reactions in Trickle Bed Reactors   163
  Summary   165
  References   165

5. Reactor Performance and Scale-Up   171
  Introduction   171
  Reactor Performance   173
    Effective Reaction Rate and Performance   173
    Particle Characteristics   180
    Gas–Liquid Distributor   182
    Liquid Maldistribution and Performance   185
    Residence Time Distribution   190
    Periodic Operation and Performance   192
  Trickle Bed Reactor Design and Scale-Up   195
    Reactor Scale-Up/Scale-Down   195
    Reactor Scale-Up Methodologies   197
    Reactor Parameters, Scale-Up, and Performance   198
  Engineering of Trickle Bed Reactors   203
  Summary   206
  References   207

6. Applications and Recent Developments   211
  Introduction   211
  Examples of Trickle Bed Reactor Applications   212

　　　Hydrogenation Reactions　　　　　　　　　　　　　　212
　　　Hydroprocessing Reactions　　　　　　　　　　　　224
　　　Oxidation Reactions　　　　　　　　　　　　　　　231
　Recent Developments　　　　　　　　　　　　　　235
　　　Monolith Reactors　　　　　　　　　　　　　　　235
　　　Micro-trickle Bed Reactors　　　　　　　　　　　247
　Closure　　　　　　　　　　　　　　　　　　　251
　References　　　　　　　　　　　　　　　　253

Notations　　　　　　　　　　　　　　　　　　　　257
Author Index　　　　　　　　　　　　　　　　　265
Subject Index　　　　　　　　　　　　　　　　271

Preface

Trickle bed reactors are widely used in chemical and petroleum processes due to their unique advantages in handling of catalysts on a large scale and operation at high pressures over the slurry reactors. The overall performance of trickle bed reactors depends on several issues that influence hydrodynamics, fluid phase mixing, interphase and intraparticle heat and mass transfer and reaction kinetics. Some specific issues like particle shape and size distribution, bed packing characteristics, flow maldistribution, wetting of catalyst particles and their influence on heat and mass transfer rates are critical for designing industrial trickle bed reactors. The innovation and competitive edge of any technology based on trickle bed reactors therefore rests on fundamental understanding and controlling the underlying fluid dynamics to suit the specific requirements and their coupled influence with reaction kinetics. Despite several conceptual advances and commercial success stories of trickle bed reactors in industry, state-of-the-art computational models are still not used for the design and simulation of trickle bed reactors. This book is aimed at filling this gap and providing a detailed account of the methodology of reactor engineering of trickle bed reactors by combining the conventional reaction engineering models with the computational flow models. The intended users of this book are chemical engineers working in chemical and petroleum industries, industrial R&D laboratories as well as chemical engineering scientists/research students working in the field of reactor engineering. Some prior background of reaction engineering and numerical techniques is assumed.

We have been involved in a variety of fundamental research and consultancy projects in the area of reactor engineering (at National Chemical Laboratory, University of Kansas, and Tridiagonal Solutions). Through our interactions with practicing engineers and research scientists dealing with trickle bed reactors, we realized that there is need to use recent advances in physics of multiphase flow along with the state of the art computational models to reactor engineering applications. We have written this book with an intention to describe the individual aspects of trickle bed reactor engineering in a coherent fashion that may be useful to further improve the design methodologies and optimize reactor performance. The information in the book is organized in a way to facilitate the task of a reactor engineer to relate reactor hardware to its overall performance. Need for using different modeling approaches for addressing different reactor engineering issues is discussed. The necessity of establishing clear relationship between reactor engineering and computational modeling is emphasized with the help of practical examples and case studies. The selection of these examples may seem to be somewhat biased since many of these are drawn from our own research and consulting experience. However, we have made an attempt to evolve general guidelines, which will be useful for solving practical problems and particularly on understanding the scale-up issues. Some comments on the recent developments in trickle bed reactors are also included.

The material included in this book may be used in several ways and at various stages of designing of trickle bed reactors as well as of the fundamental research projects on understanding the reactor performance. It may also be used as a basic resource of methodologies for appropriate reactor engineering analysis, estimation of design parameters and making decisions in practice. The content could be useful as a study material for an in-house course on reactor design, operation and optimization of trickle bed reactors or a companion book while solving practical reactor engineering problems. We hope that this book will encourage chemical engineers to exploit the potential of computational models for better engineering of trickle bed reactors.

We are grateful to many of our associates and collaborators with whom we worked on different industrial projects. Many of our students have contributed to this book in different ways. Particularly, Mr Amit Chaudhari has read the manuscript and provided valuable comments from student's perspective. We also wish to thank the Editorial team at Elsevier for their patience and understanding during this long and arduous writing process.

Vivek V. Ranade,
Raghunath V. Chaudhari,
Prashant R. Gunjal,
October 2010

Chapter 1

Introduction

Fluids wander in the great void, some of these eject and some of these stray in voids, having found no group that they could belong to. A model and image of such wandering fluids is something we have daily before our eyes: Just look when rain fall on sand stones; you will see many tiny streams twisting and turning and moving here and there where the sunlight shows. It's as if they were in an unending conflict with squadrons coming and going in ceaseless battle, now forming groups, now scattering, and nothing lasting. From this you can imagine the agitation of these fluids in the great emptiness, so far at any rate as so small an example can give any hint of infinite events.

Modified version of De Rerum Natura

TRICKLE BED REACTORS

Trickle bed reactors are gas–liquid–solid contacting devices used in many diverse fields such as petroleum, petrochemical, fine chemicals, and biochemical industries. History of trickle bed reactors can be traced back to the eighteenth century with early applications mainly in wastewater treatment all over the world. Later on these reactors became quite popular in diverse chemical and petroleum industries because of a variety of their unique advantages for large volume processing. The worldwide capacity of materials processed via trickle bed reactors is approximately 1.6 billion metric tons/annum. The value of products processed through trickle bed reactors on an average is of the order of 300 billion US$/year (Sie & Krishna, 1998). The trickle bed reactors contribute significantly to manufacture cleaner fuels by hydrodesulfurization (HDS) and hydrodenitrogenation (HDN) reactions in refineries. Several other types of reactions involving hydrogenation, oxidation, alkylation, and chlorination are carried out advantageously in trickle bed reactors. Some examples of the industrial applications of trickle bed reactors are listed in Table 1. Current and future emphasis on cleaner fuels and increasing new applications of trickle bed reactors demand in-depth understanding of the engineering tools to enable manipulation and control of the trickle bed reactors for improved performance. In recent years, several variants of trickle bed reactors including micro reactors and mesh reactors have been introduced, while the attempts to further develop the design and scale-up methodologies continue. It is therefore essential to clearly understand the

Trickle Bed Reactors. DOI: 10.1016/B978-0-444-52738-7.10001-4

TABLE 1 Some Applications of Trickle Bed Reactors

Reaction Type	Process	Catalyst	Pressure (MPa)	Temperature (K)
Oxidation reactions	Ethanol oxidation	Pd/Al	2	343–373
	Wet oxidation of phenol	Pt/Al$_2$O$_3$	3–10	100–200
	Oxidation of formic acid/oxidation of organic matter in waste water treatment/oxidation of phenol	Co/SiO$_2$–AlO$_2$, CuO	0.1–1.5	300–403
Petroleum processing	Hydrodesulfurization Hydrodenitrification Hydrodemetallization	Mo–Ni	20–80	593–653
	Hydrodemetallization			
	Catalytic hydrocracking/catalytic hydrofinishing			
	Catalytic dewaxing, dearomatization			
Hydrogenation reactions	Hydrogenation of various petroleum fractions, nitrocompounds, carbonyl compounds, carboxylic acids to alcohols (adipic acid to 1,6-hexanediol)	Pd, Pt, Ni, Cu	3–10	323–423

		Catalyst	Pressure	Temperature
	Selective hydrogenation of acetylene to separate compound from C4 fraction in the presence of butadiene	Au/Al, Pd/Al$_2$O$_3$	0.1–2.5	313–523
	Hydrogenation of crotonaldehyde and α-methylstyrene to cumene	0.05% Pd on Al$_2$O$_3$	0.1–5	373–773
	Hydrogenation of 2-butyne-1,4-diol	Ni	10–30	350–450
	Hydrogenation of caprolactone and adipic acid	Cu	15–25	450–550
	Hydrogenation of aniline to cyclohexylaniline	Pd/Al$_2$O$_3$	3–20	298–313
	Hydrogenation of glucose to sorbitol	Ru/C	8	373–393
	Hydrogenation of maleic anhydride	Raney nickel, Pt/C	1–5	200–400
	Hydrogenation of acid esters to alcohols			
	Hydrogenation of coal liquefaction extracts	Ni–Mo/Al$_2$O$_3$	7	593–623
Esterification	Esterification of acetone and butanol	Strong acidic ion exchange resin		
F–T synthesis	Fischer–Tropsch reaction	Co/TiO$_2$	10–50	450–650

fundamentals of contacting gas and liquid reactants in "trickle bed" type of reactors to realize their full potential for existing and emerging applications. This book provides the basic understanding and a computational framework to simulate, manipulate, and control the performance of trickle bed reactors.

Trickle bed reactors comprise a family of reactors in which gas and liquid phase reactants flow in downward direction (toward the direction of gravity) over a bed of solid catalyst particles. The gas phase may flow in upward or downward direction depending on the type of application. The liquid phase, however, always flows or "trickles" over the solid catalyst in a downward direction. The word "TRICKLE" itself describes its operational characteristics in which liquid intermittently flows over the solid catalyst in the form of films or rivulets or droplets. Conventionally, solid catalyst particles (may be of different shapes) are randomly packed in a bed through which gas and liquid phases flow. However, different variants of trickle bed reactors, comprising structured catalytic beds like monoliths in which the catalyst is coated on the interior surface of small channels or a mesh reactor in which the catalyst is coated on the mesh, have been used. In all these cases, catalyst loading in the reactor is substantially higher than other multiphase reactors (solid volume fraction is usually larger than 0.4 in trickle bed reactors). In most of the industrial trickle bed reactors, catalyst particles are generally porous and are of different shapes such as spherical, cylindrical, extrudates, trilobes, or multi-lobes (Fig. 1). The reactions carried out in trickle bed reactors are often exothermic and energy liberated because of chemical reactions is transported by flowing gas and liquid components. Management of this liberated energy from the catalytic bed without causing undesired effects on the performance is often a crucial task in the design of trickle bed reactors.

The overall performance of trickle bed reactors depends on several issues like characteristics of catalytic bed (packing configuration, porosity, particle size/coating thickness), flow maldistribution, wetting of catalyst particles and local interphase heat (including beds to wall) and mass transfer rates, intraparticle mass and heat transfer, and reaction kinetics. The configuration and character-istics of the catalytic bed influence the underlying fluid dynamics and therefore the local transport rates, wetting, and mixing of fluid phases in trickle bed reactors. The innovation and competitive edge of any technology based on trickle bed reactors therefore rests on how well the underlying fluid dynamics is understood and optimized to suit the specific process requirements. The complex

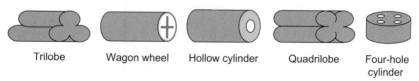

Trilobe   Wagon wheel   Hollow cylinder   Quadrilobe   Four-hole cylinder

FIGURE 1 Schematic shapes of catalyst particles used in practice (*from Palmisano, Ramachandran, Balakrishnan, & Al-Dahhan, 2003*).

fluid dynamics of trickle bed reactors often makes the scale-up or scale-down of trickle bed reactors quite difficult. Despite knowing this for several years, conventional methods used for design and optimization of trickle bed reactors often rely on experiments and empirical models. However, experimental measurements of design parameters have limitations due to severe operating conditions (high pressure and temperature) and difficulties because of the opaque and inaccessible nature of the packed beds, especially on pilot or large-scale reactors. The correlations and basic reaction engineering models based on mass and energy balances for designing of trickle bed reactors have been discussed extensively in a review by Satterfield (1975) and in a classic book by Ramachandran and Chaudhari (1983). In the last couple of decades, extensive research on various aspects of the trickle bed reactors and new developments in experimental as well as theoretical approaches has been carried out to get better insight into the complexities of trickle bed reactors. For example, new experimental techniques like Computed Tomography (CT) and Magnetic Resonance Imaging (MRI) can provide more details on porosity and gas—liquid distribution inside the trickle bed reactor at realistic operating conditions. Recent advances in computational resources and numerical techniques based on computational flow modeling now allow simulation of gas—liquid flow and phase distribution inside the column. These efforts are useful for reducing pilot plant experiments and minimizing empiricism in design and scale-up. This book provides an up to date and state-of-the-art framework for understanding, designing, and optimizing trickle bed reactor performance.

In sub-section, *Basic Configurations and Operation of Trickle Beds*, aspects of trickle bed reactor operation are discussed along with a brief comparison with other multiphase reactors, and major applications. In the subsequent section, reactor engineering of trickle bed reactors (key issues, conventional methods, and role of computational flow modeling) is discussed. In the last section of this chapter, organization of the book is outlined.

Basic Configurations and Operation of Trickle Beds

Trickle bed reactor configurations can be broadly classified into the following three types:

a. *Conventional trickle bed reactors*: comprising randomly packed beds of porous catalyst particles.

b. *Semi-structured trickle bed reactors*: comprising non-randomly packed particles or catalyst coated on structured packing [like SULZER KATA-PACK] or monolith reactors comprising large number of small channels coated with a catalyst layer.

c. *Micro-trickle bed reactor*: comprising a number of micro-channels packed with catalyst particles.

These three types are schematically shown in Figs. 2—4, respectively.

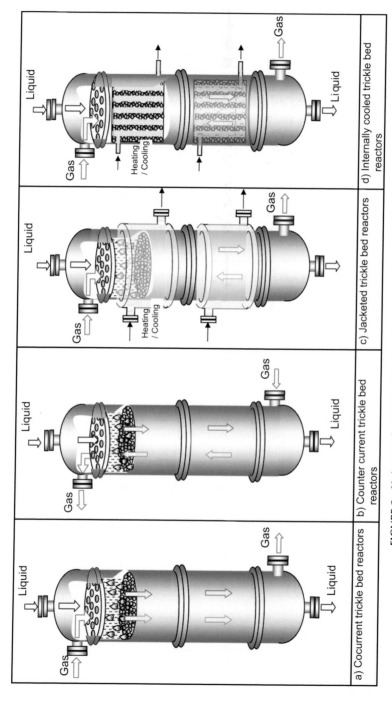

FIGURE 2   Various configurations of trickle bed reactors based on the type of operation.

a) Cocurrent trickle bed reactors

b) Counter current trickle bed reactors

c) Jacketed trickle bed reactors

d) Internally cooled trickle bed reactors

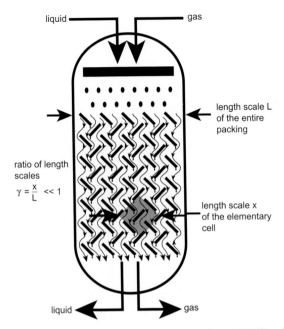

FIGURE 3   Structured trickle bed reactor *(Mewes, Loser, & Millies, 1999).*

In trickle bed reactors, gas and liquid phase reactants flow in downward direction over a bed of solid catalyst particles (Fig. 2a). In some cases, the gas phase may flow in upward direction (e.g., for specific applications like pollution control and removal of gas phase impurities), when reactor is operated in a countercurrent mode (Fig. 2b). In conventional trickle bed reactors, catalyst supported on inert material is used to provide adequate mechanical strength to the pellets. Some catalysts are in the form of eggshell where the outer layer is impregnated with an active catalytic material on a core region made up of inert support. This avoids high temperature gradients inside the catalyst particle when reactions are highly exothermic. Various catalyst shapes like spherical, cylindrical, extrudates, and trilobes are used in practice. In some cases, particle shapes such as cylindrical tubes, Raschig rings and wire gauge, pall rings, and filaments, which give lower pressure drop (at the cost of lower catalyst loading), are used. The catalyst bed is usually supported on a sieve plate (with wire mesh). The gas and liquid phases are fed to the reactor bed via appropriate distributor. Proper distribution of the fluid phases in the catalyst bed often controls the reactor performance and heat transfer efficiency. Liquid distributors are generally in the form of multiple dividing nozzles with opening at various radial positions with central inlet. Other types like bubble cap distributor, sieve plate distributor, or layer of fine particles are also used at the top of the column for achieving uniform distribution. In some large trickle bed reactors, redistribution of reactant phases is necessary to avoid hot spot

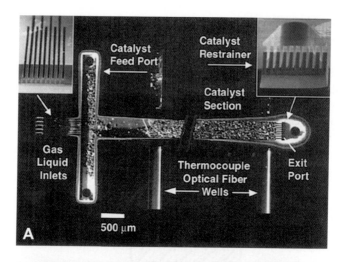

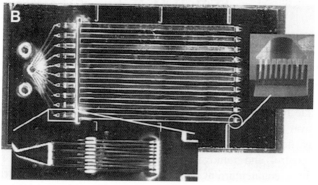

FIGURE 4 Packed bed micro-reactors. (A) Photomicrograph of packed bed reactor with active carbon catalyst. (B) Multi-channel reactor *(Losey, Schmidt, & Jensen, 2001).*

formation inside the reactors. Controlling bed temperature is one of the major concerns in trickle bed reactors. It can be performed by intermediate quenching using external jackets or internal cooling coils (Fig. 2c and d). In some cases, the gas and/or liquid streams are recycled to increase effective fluid velocity to control temperature and also manipulate desired conversion levels. Unconverted reactants and products formed are taken out from the bottom of the reactors. The bottom portion therefore consists of a gas–liquid separator. For large volume processing, multiple reactors may be operated in series or parallel. For some fine and specialty chemical manufacturing trickle bed reactors are also operated in a semi-batch mode (liquid as a batch with complete recycle).

Whenever requirement of catalyst loading is not high or mechanical strength of the catalyst is not very good, non-random packing can be used. These structured packings require lower pressure drop to operate (see Fig. 3).

The structured packings of different varieties include coated structured packing or monolith channels (structured beds). Other operating features and possibility of using intermediate quenching and redistribution are also applicable to these reactors. Some monolith reactors comprise just a single monolith so that liquid maldistribution along the length of the bed can be avoided. However, liquid distribution at the inlet becomes very critical in this case. Liquid distribution, wetting of the channel surface, and possibility of drying out of the surface because of vaporization caused by energy liberated due to chemical reactions are some of the important concerns in monolithic reactors.

In recent years, other versions of trickle bed reactors have been proposed based on the concept of micro-channel reactors (Fig. 4). Reactor functionality is similar to trickle bed reactors, however, size of the reactor is several orders of magnitude smaller than the conventional trickle bed reactors. It comprises several small channels in which catalyst is either impregnated over the wall or packed in the form of even smaller particles. Gas and liquid phases are passed using suitable distribution arrangement and alternate fins may be allocated for cooling the bed. These reactors are very compact in size and higher mass and heat transfer rates can be achievable. Therefore, these types of reactors are generally recommended where the reaction is very rapid and exothermic with significant mass transfer limitations. These reactors are also useful to control temperatures, especially when the reactant/products are sensitive to heat. Initial cost and operational difficulties (clogging, cleaning) of micro-trickle bed reactors are some of the disadvantages over the conventional trickle bed reactors.

Comparison with Other Reactors and Applications

Gas—liquid—solid contacting can be practiced in stirred slurry reactors, ejector loop reactors, bubble column slurry reactors or three-phase fluidized bed reactors, packed bubble column reactors, and trickle bed reactors. In three-phase stirred reactors, one or more rotating impellers are used to realize intimate contacting of gas—liquid—solid phases. The solid catalyst particles used in such reactors are usually much smaller than those used in trickle bed reactors. The three-phase slurry stirred reactors offer significant degrees of freedom to a designer by offering a wide range of impellers and operating speeds. However, constraints on solid loading in such reactors and mechanical difficulties associated with the moving parts (volume of reactors, mechanical seals, and maintenance) often limit the applications of stirred slurry reactors to relatively faster reaction systems requiring lower catalyst loading, low pressure operations and medium volume applications such as fine and specialty chemicals. Constraints on solid loading also apply to ejector loop reactors in which gas—liquid—solid contacting is achieved using specially designed ejector and diffuser assembly. In these reactors, slurry is circulated using a high pressure slurry pump through the ejector and reactors are usually operated as dead-end reactors. Ejector loop reactors offer flexibility of using external heat exchanger and therefore are useful to handle highly exothermic reactions. These reactors are used for even faster reactions and are therefore limited to handling limited solid loading.

Bubble column slurry reactors and three-phase fluidized bed reactors offer capability to handle relatively higher solid loading. These reactors are simple in construction, contain no moving parts, and may handle solid volume fractions up to 20–30%. These reactors offer excellent contacting and mass and heat transfer characteristics. Liquid phase residence time can be manipulated and therefore these reactors are suitable for slower reactions requiring larger catalyst loading. However, there is a significant back mixing in these reactors, which may cause lower conversion and possibility of formation of side products. For slower reactions requiring larger solid loading (volume fraction more than 30%), it is essential to use either packed bubble columns or trickle bed reactors. Slurry reactors also have limitations of substrate to catalyst ratio compared to packed bed reactors.

In packed bubble column reactors, gas and liquid phases flow through a bed of randomly packed solid catalyst particles completely filled with liquid. This reactor can be used with larger catalyst loading with gas phase flowing in the form of bubbles as a dispersed phase with liquid as a continuous phase. All the catalyst particles are always completely wetted by the liquid phase and there is no direct contact between the gas and the solid catalyst particles. Unlike this case, in a trickle bed reactor, both gas and liquid phases flow downward through the catalyst bed with liquid as a dispersed phase or semi-continuous and gas as a continuous phase. Gas phase may directly contact the catalyst surface, which under certain conditions is not completely wetted by the liquid phase. Trickle bed reactors are therefore useful for slower reactions requiring high catalyst loading and where direct contacting between gas and catalyst may benefit the overall performance. The packed bubble bed reactors usually have higher pressure drop and back mixing than trickle bed reactors.

More specific comparison of these different reactor types is given in Table 2, which indicates that trickle bed reactors are advantageous in many ways. One of the major advantages of trickle bed reactors is its simplicity in operation under high temperature and pressure conditions required for most of the industrial-scale catalytic processes. This simple operation does not require separate unit for catalyst separation and is also beneficial for lower catalyst attrition. Wide range of particle sizes can be used in trickle bed reactors (from 0.5 to 8–12 mm particle diameter). For lower particle sizes, pressure drop may be higher. On the other hand, for larger particles, reaction rates may be limited by intraparticle and interphase mass and heat transfer. However, total energy consumption is often lower because the solids are not suspended like in slurry bubble column or stirred reactor. Plug flow conditions are more favorable in trickle bed reactors compared to other three-phase flow reactors, hence higher

TABLE 2 Comparison of Different Types of Multiphase Reactors (More the Number of Stars, Better Is the Performance)

Aspect	Slurry Stirred Reactor	Ejector Loop Reactor	Slurry Bubble Column Reactor/ Three-phase Fluidized Reactor	Packed Bubble Column Reactor	Trickle Bed Reactor
Operation	*	* * *	* * * *	* * * * *	* * * * *
Solid loading	* *	* * *	* * * *	* * * * *	* * * * *
Particle size	* *	* *	* * *	* * * * *	* * * * *
Catalyst separation	*	*	* * *	* * * * *	* * * * *
Lower catalyst attrition	*	* * *	* * *	* * * *	* * * * *
Heat transfer	* * * * *	* * * * *	* * * *	* *	*
Mass transfer	* * * * *	* * * * *	* * * * *	* * *	* * *
Plug flow/lower back mixing	*	* *	* * *	* * *	* * * * *
Viscous/foaming liquid	* * * *	* * * * *	* * * *	* * * *	* *
Reactor volume	* * *	* * *	* * *	* * * *	* * * * *
Pressure	* *	* * *	* * *	* * * * *	* * * * *

conversion and selectivity is possible. However, radial temperature gradients can be significant in trickle bed reactors making it a challenge to control highly exothermic reactions in large-scale operations. In trickle beds, homogeneous side reactions are minimized due to lower liquid holdup. Similarly, higher throughput per unit volume of reactor can be achieved due to large catalyst holdup compared to slurry reactors. Due to lower liquid flow rates, partial wetting, non-uniform liquid distribution, and liquid maldistribution may lead to lower overall performance of the reactor. Partial wetting of catalyst may also favor gas phase side reactions, hotspots formation, or even temperature runaway conditions. However, these problems can be reduced using intermediate cooling, excess solvents, and liquid redistributors. Trickle bed reactors offer several advantages compared to other multiphase reactors, especially for large volume operations and are therefore chosen for wide ranging processes in chemical and petroleum industries. Some important applications of trickle bed reactors are listed in Table 1. The relevant engineering science concepts to design and scale-up of these reactors and challenges for reactor engineers are discussed in the following sections.

REACTOR ENGINEERING OF TRICKLE BED REACTORS

Trickle bed reactors are generally operated under severe conditions of high pressure and high temperature. The basic construction and operation of trickle bed reactors are rather simple as discussed in previous sections. However, such simplicity in construction and operation often leads to lower degree of freedom to manipulate and control the overall performance of such reactors. This lack of control makes the task of reactor engineering quite complex. Reactor engineering of trickle bed reactors therefore needs to take into account all the processes occurring on various spatial and temporal scales and understand them to realize the desired performance in practice. Typical processes occurring in trickle bed reactors are briefly discussed in the following.

When reactants in gas and liquid phases are passed through a catalytic bed, several processes occurring on macro-scale, meso-scale, and micro-scale control the overall rates of chemical reactions involved in the process. Depending on the degree of wetting, gaseous reactants reach the catalyst surface either via liquid phase or by direct contact. The intraparticle diffusion of dissolved gas and liquid reactants, adsorption on catalyst surface, and chemical reactions occur simultaneously in catalyst particles. The products diffuse back from the sites through intraparticle diffusion to the bulk liquid and gas phases. The concentration of gas and liquid phase reactants in the bulk fluid phases is influenced by the extent of back mixing in these phases. In the case of liquid phase, there is an additional complexity of existence of stagnant liquid pockets between catalyst particles. The liquid volume fraction in trickle bed reactors is usually characterized as dynamic liquid holdup and static liquid holdup (corresponding to stagnant liquid pockets). The exchange between these two quantities often determines the effective residence time distribution in trickle bed reactors. It should also be noted that configuration of particles affects the flow within the bed and controls the overall pressure drop, liquid volume fractions (static and dynamic), and therefore effective wetting, heat, and mass transfer rates. The overall gas and liquid phase flow rates determine the prevailing flow regimes. It should be noted that different flow regimes may exist in trickle bed reactors with very different contacting and mixing characteristics. Energy liberated due to chemical reactions further complicates the picture by possible evaporation and drying out of the pores, leading even to internal drying of catalyst particles. All such issues need to be carefully addressed for systematic design of trickle bed reactors, a brief outline of which is given in the following section. The details of methodology to quantitatively address these issues are discussed in subsequent chapters of this book.

Key Issues

In trickle bed reactors, different transport processes occur at different time and length scales, a schematic of which is shown in Fig. 5.

On a reactor scale, gas and liquid phases are introduced from the top, which flow through the voids of catalyst bed. Several different flow regimes, like trickle flow, pulsing flow, and spray flow regimes, may exist in trickle bed reactors because of different levels of interphase interactions. These macroscopic flow regimes (occurrence, regime maps, and stability) and their influence on other transport characteristics need to be understood as primary issues in reactor engineering analysis of trickle bed reactors. Physical properties of gas and liquid phases (and their variation with pressure and temperature), packing characteristics (size, configuration, wetting characteristics, etc.), distributor designs, reactor size, and flow rates of gas and liquid phases influence prevailing flow regimes in any given trickle bed operation. Study of flow regimes and overall reactor performance is obviously related to macroscopic scale and require modeling of an entire reactor system.

For any prevailing flow regime, the distribution of liquid phase reactants in the bed depends on quality of liquid distribution at the inlet, overall variation in bed porosity, wetting, and capillary forces. Design of gas–liquid distributor, location of the distributor, and top layer of the packing are important parameters to achieve uniform distribution and to increase effective utilization of the catalyst bed. Though liquid is distributed uniformly in the top region of the column, non-uniformities in bed porosity and uneven wetting may cause further non-uniformities as the liquid flows along the length of the reactor. Packing configuration significantly influences possible channeling and maldistribution within the reactor. Local particle configuration and therefore local variation of porosity is dependent on the type of packing used, size and shape of the particles, and ratio of particle to column diameter and so on. Local gradients in porosity influence the capillary forces and therefore divert the flow path affecting wetting of catalyst particles. It is important to understand and estimate whether redistribution of fluid phases is necessary to ensure the uniformity in distribution, especially in tall columns.

Different local packing arrangements exist in randomly packed bed reactors and nature of voids formed between particles affects the flow structure inside the void and hence controls the mixing, heat, and mass transport rates. It also affects the stagnant and dynamic liquid holdup in the bed. For exothermic reactions, this phenomenon is important where dry out of particles may lead to the formation of local hotspots. Local hot spots may lead to temperature runaway, posing safety issues. Solvent evaporation adds further complexities in heat and mass transfer rates, most often proving beneficial for temperature control. The analysis of these processes requires careful account of flow, mixing, and other transport processes occurring around a cluster of particles comprising typically tens of particles. Processes

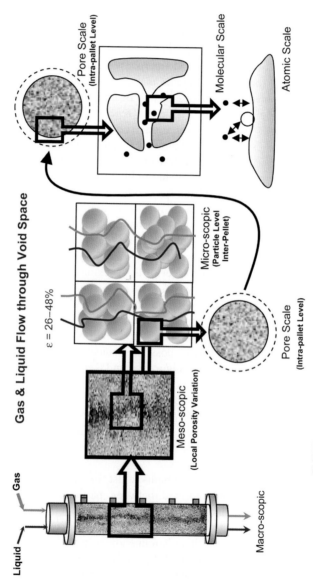

FIGURE 5 Schematic of various process scales involved in trickle bed reactors.

occurring on these scales are typically called as meso scale processes. In randomly packed bed, there is a limited scope to alter the local characteristics of packing/voids for manipulating the local mixing, and transport characteristics. The manipulation of particle size and shape (spherical, cylindrical, trilobes, and so on) may allow some degree of control. Use of structured packing in trickle bed reactors has been developed to obtain better control on local packing characteristics and therefore on local mixing and transport rates.

On a single catalyst particle scale, wetting of particle and intraparticle mass and heat transfer play an important role in the overall rate. Though wetting of particle is a result of global operations (liquid distribution, liquid flow rates), particle-scale parameters, e.g., capillary forces, particle shape, contact angle, surface characteristics, and local gas and liquid flows, also determine the degree of wetting of each particle. Flowing liquid forms a film over the external surface of the catalyst and partial wetting may occur at lower liquid flow rates. Wetted part of the catalyst surface gets exposed to the liquid phase reactants and the dissolved gas phase reactants while the non-wetted part is exposed only to the gas phase reactant. In most cases, however, due to capillary effects, catalyst particles get completely filled with liquid phase, and hence at the surface of even the unwetted part of the catalyst both liquid and gas phase reactants exist. However, this condition is not always true if liquid phase is evaporating or pores are larger such that capillary effects are negligible. Partial wetting condition affects the reaction rates in various ways depending upon the reaction conditions and the properties of the fluids. For kinetically controlled reactions, reaction rates are directly proportional to the extent of internal wetting of particles. Partial wetting condition influences mass transfer from gas and liquid phases to catalytic sites available on particle (outer surface and internal pores). Gas–particle mass transfer rates are significantly enhanced due to direct access of gaseous reactants through the non-wetted surface. Analysis of reaction rates under partial wetting condition is extremely complex due to solvent and substrate condensation/evaporation, local temperature variation, and hence the rates of diffusion, adsorption, and reaction. Many times this leads to an increase/decrease of catalyst effectiveness and in some cases to multiplicity of conversion and temperature. It is therefore often useful to develop single particle models to get an insight into local processes occurring in a trickle bed reactor. Most of the trickle bed reactions are exothermic in nature and careful account of intraparticle, interphase, and even bed-to-wall heat transfer is crucial in understanding the overall performance.

On a scale smaller than catalyst particle, it is important to understand the thermodynamics of adsorption and mechanism of chemical reactions taking place on catalyst sites. There can be more than one type of catalyst sites and sites may transform from one type to the other. Understanding of the interaction of sites and adsorbed species and mechanism of chemical transformations using a micro-kinetic modeling approach helps in developing realistic kinetic models which can be used to optimize overall performance of reactors. This will require linkage of the kinetic model with appropriate single particle, meso-scale, and macro-scale reactor engineering models.

The conventional methods (experimental and semi-empirical methods) used to model trickle bed reactors usually provide only an overall and global description. This practice conceals detailed local information about the mixing and transport parameters which may ultimately determine the reactor performance. Considering the interaction of processes occurring on a wide range of spatio-temporal scales and complexity of flow in trickle bed reactors, it is essential to develop a comprehensive reactor engineering approach by harnessing recent advances in computational modeling. In recent years, chemical engineers have exploited the power of computational flow modeling for reactor engineering applications. Many chemical industries have initiated efforts in this direction. However, still many chemical engineers either consider that the flow complexities of industrial reactors are impossible to simulate or expect miracles from off the shelf, commercial flow modeling tools. These two diverse views arise because there is not much interaction between the flow modeling and the industrial reactor engineering communities. This is especially true for the case of trickle bed reactors. With the emergence of high performance computers and advances in physics of multiphase flows and numerical techniques/algorithms, it is now possible to develop and use a multiscale modeling strategy for reactor engineering of trickle beds.

Multiscale Approach for Reactor Engineering of Trickle Beds

Conventional reactor engineering of trickle bed reactors is mostly based on simplified reaction engineering analysis which involves overall material and energy balance. Averaged properties and empirically evaluated model parameters are generally used in these models. Empirical correlations known to date are mostly evaluated through cold flow experiments and laboratory or pilot-scale reactors. These correlations may not be valid for industrial reactors because hydrodynamic and flow characteristics in these may be quite different than the laboratory or pilot-scale reactors. Primarily, the uncertainties associated with such models and designs based on them are related to lumped descriptions of the processes occurring on widely different spatio-temporal scales. Various processes occurring on different scales (from molecular scale to tens of meters) interact with each other. Such interactions over a wide range of scales cause severe difficulties in developing and solving predictive models to simulate reactor performance. It is often difficult to develop a single model to describe a complete reactor performance. Therefore, it is necessary to develop multilayer or multiscale models, which comprise different models, each simulating processes occurring on different scales with different overall objectives and level of complexity. These models communicate information with each other to provide adequately accurate estimation of the overall performance. Such an approach for reactor engineering of trickle bed reactors is discussed in this book.

Multiscales are inherent in nature. Therefore developing multiscale modeling approaches is gaining increasing interest in many engineering sciences including reactor engineering. Real life processes may span over a wide range of length and time scales (several orders of magnitude). Modeling and resolving all these scales spanning over a wide range may not be feasible and necessary. For the trickle bed reactors, various processes of interest may be divided into three groups as:

- micro-scale processes (events on catalytic sites and their relations to thermodynamics, reaction kinetics, and transport steps)
- meso-scale processes (events occurring at clusters of particles), and
- macro-scale processes (events occurring at bed scale)

In many cases, models based on macro-scales are sufficient for process design and have been used successfully in solving many practical problems. However, models without considering micro-scale and meso-scale processes may lead to inaccurate solutions and severely limit generalization of models. Solutions of micro-scale and meso-scale processes may provide significant insight into process complexities and may enhance reliability of the models by reducing empiricism. The micro-scale and meso-scale process modeling, however, is often computationally intensive and provides excessive information compared to that usually required for the process design. Therefore, one of the major tasks in multiscale modeling is to judiciously combine micro-scale, meso-scale, and macro-scale models to achieve the engineering objectives (enhance the utility, reliability, and effectiveness without compromising tractability). Many techniques such as model reduction methods, domain decomposition methods, models based on self-similarity, scale separation methods, multigrid techniques, and adaptive mesh refinement may be useful for this. Application of multiscale modeling methodology for trickle bed reactors requires thorough understanding of the processes occurring inside trickle bed reactors (see Fig. 5). Multiscale modeling approach applied to these different processes in trickle beds is schematically shown in Fig. 6. Different transport/ transformation processes occurring at various scales (ranging from pico-scales to mega-scales) in trickle bed reactors are shown in small lime-colored templates while plausible modeling approaches are shown in larger cyan-colored templates. These are briefly discussed in the following.

- The smallest scale for which any design engineer would be interested is the pico-scale, where interaction among individual atoms of different species and phases occurs. Physical phenomena at this scale comprise movement of atoms controlled by the laws of the thermodynamics and chemical potential, occupation of catalyst active sites, and chemical kinetics for the reacting system. Quantum mechanical calculations (based on Density Functional

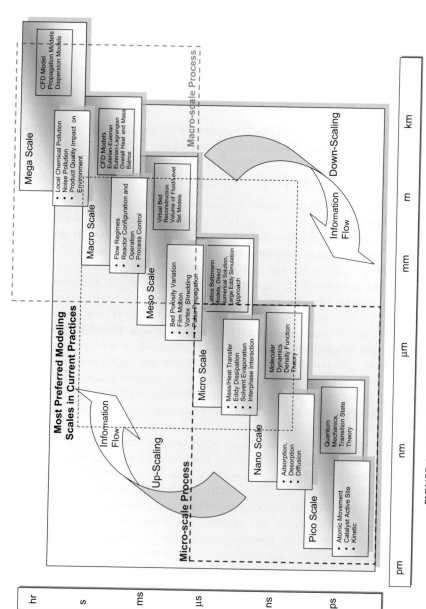

FIGURE 6 Multiscale phenomena and modeling approaches for trickle bed reactor.

Theory, for example) are useful for simulating such types of atomic interactions. Potential surface energies and activation energies can be calculated using these models, which are useful for the next level of modeling at nano-scales. Evaluation of chemical kinetics is theoretically possible using such an approach.

- Nano-scale phenomena comprise interaction among several molecules among each other and with the surface of the catalyst within the catalyst pore. Physical phenomena occurring within the domain of this scale is adsorption, desorption of liquid and gas phase species on catalyst active site, catalyst deactivation, and diffusion of species within pores. This scale is mostly confined to the phenomena occurring on a catalyst surface at a pore or pore mouth level. Approaches used for modeling these phenomena are based on molecular dynamic calculations, DFT approach, and Monte Carlo simulations. These simulations are useful for finding out activation, deactivation mechanisms, and net flux into the particle.

- Micro-scale processes are generally associated with a single particle and their surroundings. Key processes occurring at this scale are energy exchange between the phases, heat and mass transfer across the flowing fluids and solid surfaces, and unique features like incomplete wetting of catalyst particles. The analysis at this scale is possible using advanced modeling techniques such as lattice Boltzmann models or large eddy simulations for porous media. For partially wetted particles, solvent evaporation and capillarity phenomenon occurring inside the particle are associated with this scale. Volume of fluid approach or front tracking methods may be used for understanding partial wetting phenomena over a particle.

- Meso-scale processes are related to flow and transport (heat and mass transfer) around group of particles. Fluid flows through the voids formed by these particles and takes tortuous path through these complex geometries. When particles are randomly packed, local variation of porosity causes uneven distribution of flowing fluids (local maldistribution). Size and shape of particles, orientation of particles, fluid properties, and their flow rates govern these meso-scale processes. When gas–liquid flows through confined space and flow rates are moderate to high, pulse formation and propagation occurs which leads to a very different interaction than trickle flow. Meso-scale models may provide insights into pulse formation and other key issues of gas–liquid flows through complex voids. Reconstruction of void distribution to build a virtual porous media is a major task for developing suitable computational model. Surface wetting and spreading of liquid can be modeled using volume of fluid approach or other front tracking methods like level set methods.

- Macro-scale processes are associated with the reactor scale. The models for this scale may range from simplified reaction engineering models (like plug flow reactor model) to complex three-dimensional computational fluid dynamics (CFD)-based models. Classical reactor engineering models are based on idealizing mixing behavior of fluids inside the reactors (for example, plug flow or completely mixed reactor models). Intermediate or non-ideal mixing behavior can be modeled using dispersion models or multizonal models. The CFD models can account for detailed reactor configurations, internals, and structure of packing, cooling/heating arrangements. The macro-scale models, however, require closure models to account for processes occurring on meso-scale and micro-scale. If appropriate closures are used, macro-scale models are capable of predicting mixing in reactors, possible maldistribution, as well as effect of internals on performance of reactors. Such models are useful for optimization of operating parameters and developing better control strategies.

- Purpose of mega-scale models is different from the models which are discussed earlier. Primary aim of the modeling approaches discussed above is to study the impact of process parameters on the performance of a reactor. Purpose of mega-scale modeling is to study the impact of reactor performance on overall plant and on surrounding environment. Domain of processes at this scale may involve local plant area, surrounding metropolis, and even entire global environment. Physical processes associated on this scale are pollution due to unwanted products formation (waste), noise generated due to reactor operation, reactor performance-dependent pollution (solid, liquid, and gaseous), and so on. Different kinds of models are used for this purpose, which are beyond the scope of this book.

In this book, the scope is limited to the discussion of micro-, meso- and macro-scale processes (regions surrounded by black dotted lines in Fig. 6). The organization of the book is discussed in the following section.

ORGANIZATION OF THIS BOOK

We have written this book with an intention of explaining and relating the individual aspects of reactor engineering of trickle beds in a coherent way. The information in the book is organized in a way to facilitate the central task of reactor engineer, i.e., relating reactor hardware to reactor performance. Several steps to achieve such a task are discussed to clearly identify the interaction among various issues. Need for using different modeling approaches for addressing various reactor engineering issues are discussed. A methodology to synthesize results from different modeling studies to realize better reactor engineering is discussed. The necessity of establishing a clear relationship between objectives of reactor engineering and objectives of computational modeling is emphasized. Several examples and case studies covering three major applications viz, hydrogenation, oxidation (including wasterwater treatment), and hydroprocessing are discussed. The selection of these examples may seem somewhat biased since many of these are drawn from our research and consulting experience. However, we have made an attempt to evolve general guidelines, which will be useful for solving practical reactor engineering problems. Some comments on future trends in enhancing performance of trickle bed reactors are also included.

This book consists of six chapters, which provide a brief introduction to key design and operating characteristics of trickle bed reactors and differentiate these from other gas—liquid—solid reactors. Critical aspects of reactor engineering of trickle beds are discussed. A need for a comprehensive multiscale modeling approach for reactor engineering of trickle bed reactors is highlighted. At the end of this part, overall organization of the book is explained.

In Chapter 2, hydrodynamics and flow regimes in trickle bed reactors are discussed in detail. Different flow regime characteristics in the presence of single and multiphase flows through packed bed are thoroughly discussed. Some of the theoretical as well as experimental flow regime identification techniques are presented with a few illustrative results from the literature. Accurate estimation of hydrodynamics parameters such as pressure drop, liquid holdup, wetting efficiency, and mixing is required for designing trickle bed reactors. Available information is critically analyzed to evolve guidelines for the estimation of model parameters. Estimation of heat and mass transfer rates (gas—liquid, liquid—solid, and gas—solid) is an important aspect of trickle bed reactor design. Key correlations for estimating these parameters are discussed in this chapter.

Chapter 3 discusses the reaction engineering models for trickle bed reactors along with applications of these models for designing trickle bed reactors. This covers methods of reaction rate analysis, basic mass, and energy conservation equations to describe operation of trickle bed reactors and models for some of the special cases of operation of trickle beds. The particle level rate analysis incorporating the contributions of reaction kinetics, external mass transfer, and intraparticle diffusion processes has been described for both isothermal and non-isothermal reactions and for single and multistep reactions. Integration of these fundamental issues with reactor performance models has been illustrated for a few cases including the partial wetting of catalyst particles. Basic governing equations to describe different cases like gas-limiting reaction, liquid-limiting reactions, and volatile solvent are discussed. Application of reaction engineering models in practice is discussed and possible uncertainties associated with the underlying assumptions and estimation of model parameters are highlighted.

In Chapter 4, multiscale modeling approach, computational models, and methods with a potential to address uncertainties mentioned above are discussed. The discussion on mathematical modeling of fluid flow in packed beds is divided into three main sub-parts, namely, characterization of packed bed, single phase flow, and gas—liquid flow. A need to understand and recognize the distinction between a learning model and a simulation model is emphasized. Recent attempts of understanding micro-scale, meso-scale, and macro-scale fluid dynamics in trickle beds are critically analyzed and discussed. Use of these models to enhance our understanding of fluid dynamics of trickle beds and ways of extending their application to address reactor engineering issues are discussed.

In Chapter 5, overall methodology of reactor engineering of trickle bed reactors is presented. In the first part rate analysis and performance evaluation of trickle bed reactors are discussed. Difficult issues like characterization of packed beds, influence of particle shape and size distribution, distributor design, residence time distribution, and wetting are adequately discussed. Issues of scale-up and scale-down and possibility of forced periodic operation of trickle bed reactors are discussed. Various distributors used in practice are described. Practical approaches used for design and scale-up of trickle bed reactors are discussed. Hybrid approach of using experiments and various types of models for reactor engineering is discussed. Key uncertainties, reactor engineering concerns and possible ways for mitigating these concerns about the important design and scale-up issues are highlighted.

Some applications of the methods and approaches discussed in earlier chapters are presented in Chapter 6. These case studies cover three major areas of applications: hydrogenation, hydroprocessing, oxidation, and wastewater treatment. Key aspects of these applications from the reactor engineering perspective are discussed. Selection and application of the modeling approaches for addressing these key issues are discussed. Some of the limitations of current state of knowledge in describing complex underlying physics of gas–liquid flow through trickle beds are also discussed. Suggestions are made about possibilities of using the computational models for reactor engineering applications despite their limitations. The last part covers recent advances and recent variants of trickle bed reactors, namely, monolithic reactors and micro-channel reactors.

The last section recapitulates the lessons learnt from our experience of using state-of-the-art computational models to address practical engineering problems of trickle bed reactors. Advantages of using the methodology discussed in the book and potential pit falls are reemphasized. The material presented in this section will be very useful for solving practical reactor engineering problems. It will also be useful for identifying the needs for future research on understanding trickle bed reactors.

The book is expected to impart an adequate capability to readers and equip them with a methodology with which they can design and enhance performance of trickle bed reactors used in a variety of applications.

REFERENCES

Losey, M. W., Schmidt, M. A., & Jensen, K. F. (2001). Microfabricated multiphase packed-bed reactors: characterization of mass transfer and reactions. *Industrial and Engineering Chemical Research., 40,* 2555–2562.

Mewes, D., Loser, T., & Millies, M. (1999). Modelling of two-phase flow in packings and monoliths. *Chemical Engineering Science, 54,* 4729.

Palmisano, E., Ramachandran, P. A., Balakrishnan, K., & Al-Dahhan, M. (2003). Computation of effectiveness factors for partially wetted catalyst pellets using the method of fundamental solution. *Computers and Chemical Engineering, 27,* 1431−1444.

Ramachandran, P. A., & Chaudhari, R. V. (1983). *Three Phase Catalytic Reactors.* New York, USA: Gordon and Breach.

Satterfield, C. N. (1975). Trickle bed reactors. *AIChE Journal, 21,* 209.

Sie, S. T., & Krishna, R. (1998). Process development and scale up-III. Scale-up and scale-down of *trickle bed* processes. *Reviews in Chemical Engineering, 14,* 203.

Chapter 2

Hydrodynamics and Flow Regimes

It is futile to do with more, what can be done with less.

Occam's razor

INTRODUCTION

In Chapter 1, various technological aspects and conceptual issues relevant to reactor engineering of trickle bed reactors are discussed including the industrial applications and engineering science challenges related to design and scale-up. The importance of multiscale modeling is also discussed in the context of complexities involved in understanding the performance of trickle bed reactors at different scales. It is, however, important and essential to use the accumulated knowledge and wisdom of operating trickle beds to complement and to realize the real benefits of the multiscale modeling. Significant body of experimental data and models are available on various aspects of hydrodynamics, mass and heat transfer, and reactor engineering of trickle bed reactors. Available knowledge on hydrodynamics, flow regimes, and transport parameters in the trickle bed is critically reviewed and presented in a usable form in this chapter. The next chapter (Chapter 3) describes conventional reaction engineering models for trickle bed reactors.

Reaction engineering models (discussed in Chapter 3) based on overall material and energy balances require various design parameters representing hydrodynamics of trickle beds such as liquid holdup, mass and heat transfer coefficients, and axial dispersion. It is essential to establish the relationship of these hydrodynamic parameters with *the key design and operating variables* of the trickle bed reactors. These hydrodynamic parameters are usually obtained from cold flow experiments at much smaller scales than the actual scales employed in industry and the applicability of these at higher scales is often debated. Experimental data used to develop appropriate correlations for establishing relationship with the design and operating parameters are often based on model air–water system and hence their applications to other systems should be done carefully. The development of correlations based on

Trickle Bed Reactors. DOI: 10.1016/B978-0-444-52738-7.10002-6

dimensionless numbers with physical significance and on extensive experimental data has been found to be useful in many cases. This chapter provides comprehensive information on available correlations for the estimation of these parameters useful in designing a trickle bed reactor.

A trickle bed reactor can be visualized as a bed of catalyst particles with interstitial space among them forming a complex pattern of interconnecting and randomly distributed pores. When gas and liquid reactants flow over these catalyst particles, complex interactions between the flowing fluid phases and stationary solid particles lead to different flow patterns or regimes. These flow patterns essentially depend on the packing density, gas and liquid velocities, particle size, and physical properties of the fluid phases. In trickle bed reactors, at least four different flow regimes are recognized (Charpentier, Bakos, & Le Goff, 1971; Charpentier, Prost, & Le Goff, 1969; Fukushima & Kusaka, 1977a, 1977b; Herskowitz & Smith, 1983; Weekman & Myers, 1964). Earlier attempts of developing correlations for the design parameters without considering the differences in flow regimes were not very successful for extrapolation to different systems and scales. It is therefore important to understand the intricacies of different flow regimes, and the ways of identifying prevailing flow regime and transition between them for a specific configuration of the trickle bed reactor at given operating conditions. Various theoretical and experimental methods have been used to investigate different flow regimes in trickle bed reactors (see, for example, Charpentier et al., 1971, 1969; Duduković, 1977; Dudukovic, 2000; Herskowitz & Smith, 1983; Satterfield, 1975). Classification of the flow regimes and their key hydrodynamic characteristics are discussed in the following section. The transition boundaries between different regimes need to be identified accurately since more often than not trickle bed reactors are operated closer to these transition boundaries. The correlations and models for estimating flow regime transition are presented in *Flow Regime Transition* section. Several hydrodynamic and transport parameters such as pressure drop, liquid holdup, wetting of particles, and interphase mass and heat transfer coefficients and interfacial area are considerably affected by the prevailing flow regime. Different methods (models and correlations) for estimating the hydrodynamic and transport parameters in different flow regimes of trickle bed reactors are discussed in *Estimation of Key Hydrodynamic Parameters* section.

FLOW REGIMES

Packed bed reactors are composed of a pore network of complex shapes formed by randomly packed catalyst particles. When fluid is subjected to flow through such a network, it preferably flows through the path which offers least resistance. This usually results in inhomogeneous velocity and phase distribution in the bed. Different flow regimes are observed even in a single-phase flow through packed beds.

For a single-phase flow, at very low flow rates, creeping flow exists for $Re_p < 1$, where Re_p is the particle Reynolds number based on mean velocity in the void space ($Re_p = \rho U_0 d_p / \varepsilon \mu$); the inertial flow regime begins above $Re_p = 10$. This inertial flow regime extends up to $Re_p = 250-350$. With further increase in Reynolds number, the transition flow regime (unsteady flow) occurs up to $Re_p = 900$ (Seguin, Montillet, Comiti, & Huet, 1998). The transition of flow from laminar to turbulent characteristics in the packed beds is difficult to identify and occurs over a range of Reynolds numbers. The occurrence of transition to turbulence regime is a complex function of size and shape of particles and bed packing characteristics. Previous studies (for example, Chhabra, Comiti, & Machac, 2001; Seguin, Montillet, & Comiti, 1998; Seguin et al. 1998) indicate that beyond $Re_p = 350$, flow is dominated by eddies and turbulent-like structures. Seguin et al. (1998) experimentally showed that turbulent flow exists beyond $Re_p = 900$. Jolls and Hanratty (1966) and Latifi, Midoux, Storck, and Gence (1989) reported that transition occurs over the range of $Re_p = 300-400$ ($Re_p = \rho U_0 d_p / \mu$). Based on these studies, it can be assumed that flow becomes completely turbulent beyond a particle Reynolds number of 1000.

When gas and liquid phases are flowing cocurrently downward through a packed bed of solid particles (as in trickle bed reactors), the situation is much more complex compared to the single-phase flow. Most of the trickle bed reactors are operated in a cocurrent manner and hence the discussion here is mainly focused on cocurrent operation. In trickle bed reactors, various flow regimes were observed at different gas and liquid flow rates. Apart from the flow rates of gas and liquid, prevailing flow regime is also a function of parameters such as dimensions of the reactor, size and shape of particles, method of packing, and thermo-physical properties of gas and liquid phases. Broader classification of the two-phase flows in the packed bed reactors is based on the nature of flow of individual phases viz., continuous, semi-continuous, and dispersed flow. Four distinct flow regimes were identified in trickle bed reactors (Chaudhari and Ramachandran (1983):

- Trickle flow regime,
- Pulse flow regime,
- Spray flow regime, and
- Bubbly flow regime.

Names of these flow regimes indicate their typical characteristics which are shown schematically in Fig. 1. Flow characteristics of these regimes are quite distinct and therefore it is important to understand the dominating features of each flow regime. Traditionally, different flow regimes have been studied experimentally by varying either gas or liquid flow rates. At low gas and liquid flow rates, gas–liquid interaction is small and liquid flows in the form of films or rivulets over the packed particles as shown in Fig. 1a. This flow regime is known as trickle flow regime or low interaction regime. At moderate gas and

(a)

(b)

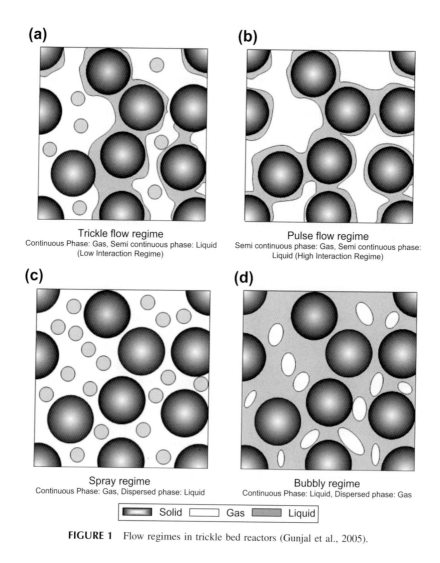

Trickle flow regime
Continuous Phase: Gas, Semi continuous phase: Liquid
(Low Interaction Regime)

Pulse flow regime
Semi continuous phase: Gas, Semi continuous phase:
Liquid (High Interaction Regime)

(c)

(d)

Spray regime
Continuous Phase: Gas, Dispersed phase: Liquid

Bubbly regime
Continuous Phase: Liquid, Dispersed phase: Gas

| ■ Solid | ☐ Gas | ▨ Liquid |

FIGURE 1 Flow regimes in trickle bed reactors (Gunjal et al., 2005).

liquid flow rates, interaction among the phases increases and liquid phase occupies entire flow cross-section. This process leads to the formation of alternate gas—liquid-enriched zones as shown in Fig. 1b and the corresponding regime is classified as pulse flow regime (also referred as high interaction regime). Trickle and pulse flow regimes occur at low-to-moderate flux of gas and liquid flow rates and industrial reactors are commonly operated in these flow regimes. Two other flow regimes (spray and bubbly) may occur at higher gas or liquid flow rates but these are less common in practical conditions. At low gas flow rate and high liquid flow rates, liquid phase occupies entire portion of the bed and becomes continuous phase while gas phase is flowing in the form of bubbles in the downward direction. This flow regime is known as bubbly flow regime and its schematic is shown in Fig. 1d. On the other hand, at low liquid and high gas flow rates, liquid phase becomes dispersed phase in the form of droplets [see Fig. 1c] and gas phase becomes the continuous phase. This flow regime is known as spray flow regime. Key characteristics of the individual flow regimes are discussed here.

Trickle Flow Regime

Trickle flow regime exists at low liquid and moderate gas flow rates where liquid flows in the form of films or rivulets over the catalyst particles. At low liquid flow rates, inertial forces are weaker compared to the local surface forces and liquid spreading is mainly controlled by capillary pressure. Hence, liquid phase flows in the form of rivulets. However, at higher flow rates, inertial forces become important as compared to the interfacial forces (adhesion, capillary); resulting in film formation over the catalyst surface. The flow regime map reported by Sie and Krishna (1998) is shown in Fig. 2 for air−water system and it is shown that the trickle flow regime exists till 12−15 kg/m^{2}s of liquid flow rates and ~1.25 kg/m^{2}s of gas flow rate. The foaming nature of the liquid phase significantly affects the flow regime boundaries. In such cases, relative importance of surface forces should be considered along with the inertial and viscous forces (see, for example, Gianetto, Baldi, & Specchia, 1970; Morsi, Midoux, & Charpentier, 1978; Talmor, 1977). Trickle flow regime region widens with increase in particle size, decrease in liquid viscosity, and surface tension. Low pressure drop, low gas−liquid throughputs, less catalyst attrition,

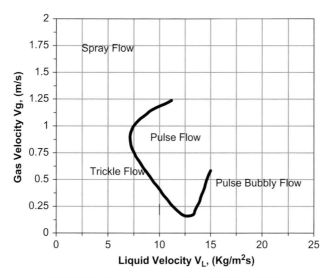

FIGURE 2   Flow regime map (Sie & Krishna, 1998).

and suitability for foaming liquids are some of the advantages of trickle flow operation. Particle wetting may be advantageous or disadvantageous depending upon the reaction type. In trickle flow regime, heat and mass transfer rates are poorer as compared to the other flow regimes. In spite of these issues, many industrial processes operate in this regime to achieve the specific process goals (like higher conversion and productivity).

Pulse Flow Regime

Transition from trickle to pulse flow regime occurs with increase in either gas flow rate or liquid flow rate. The flow regime map reported by Fukushima and Kusaka (1977a) is shown in Fig. 3. This map is in terms of gas and liquid phase Reynolds numbers. In the pulse flow regime, local flow path for gas phase is blocked by liquid pockets/plugs which results in the formation of alternate gas and liquid-enriched zones. In the liquid-enriched zones, complete wetting of particles occurs under this flow regime. Pulse flow regime boundaries are significantly affected by properties of the liquid, i.e., foaming or non-foaming. In the case of non-foaming liquid, gas—liquid-enriched zones are quite distinct and visible. Smaller-sized bubbles get entrapped into the liquid slugs. In the case of a foaming liquid, liquid slugs contain large-sized gas bubbles and the gas fraction in liquid-rich bands is also significant. In such a case, tendency of transition from the pulse flow to bubbly flow regime increases with increase in foaming nature of the liquid (Talmor, 1977). Most of the industrial trickle bed reactors are operated at close to the

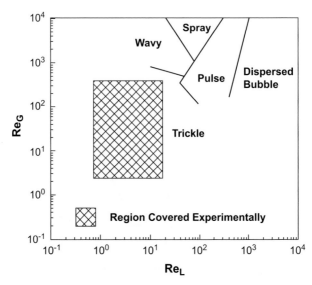

FIGURE 3   Flow regime map (Al-Dahhan & Dudukovic, 1994; Fukushima & Kusaka, 1977b).

boundary of trickle to pulse flow regime (Satterfield, 1975) taking advantages of both operating regimes.

Pulse flow regime has advantages in terms of wetting, effective utilization of catalyst bed, and higher heat and mass transfer rates. However, operating window for pulse flow regime is relatively smaller than other flow regimes such as trickle, spray, and bubbly flow regimes. Large diameter trickle beds are difficult to operate in pulse flow regime. Forced cyclic operation may be an alternative way for such cases in order to get some of the benefits associated with the pulse flow regime.

Spray Flow Regime

At high gas flow rates and low liquid flow rates, liquid film in trickle flow regime is subjected to high shear due to relatively higher slip velocity. Under certain conditions, liquid phase looses its semi-continuous nature and flows in the form of droplets. This regime is called a "spray flow" regime. It is difficult to identify and measure exact boundary between the trickle flow and the spray flow regimes. The flow regime map reported by Saroha and Nigam (1996) is shown in Fig. 4. This map is in terms of key dimensionless numbers and is quite general. Parameters λ and ψ are defined in Fig. 4. G denotes mass flux and subscripts G and L denote gas and liquid phases, respectively. Definitions of other commonly used symbols are listed in *Notations*. This flow regime map may be recommended for identifying prevailing flow regimes in trickle beds.

The spray flow regime typically occurs at $V_G > 1.25$ kg/m^{2}s and $V_L < 12$ kg/m^{2}s. Typical gas phase Reynolds number is larger than 100. The boundary between the trickle and the spray flow regimes is quite sensitive to particle diameter and surface tension as compared to other parameters. Due to higher gas flow rates, gas recycling is required if it is operated under spray flow regime. Low liquid holdup, high gas–liquid mass transfer rates, and low foamability are typical characteristics of this flow regime.

Bubbling Flow Regime

At low gas and high liquid flow rates, liquid occupies entire region of the void space in the packed bed and becomes a continuous phase. Gas flows as a dispersed phase in the form of bubbles. This flow regime is known as bubbling flow. The flow regime map shown in Fig. 4 may also be used for identifying bubbling flow regime. This way an intimate interaction among the phases is possible at the expense of higher pressure drop. Higher liquid holdup leads to back mixing which may not be suitable for some of the reactions. Complete wetting of the bed and high heat and mass transfer rates are some of the advantages of this flow regime and is suitable for cases where liquid phase is a limiting component and reactions are highly exothermic. This flow regime

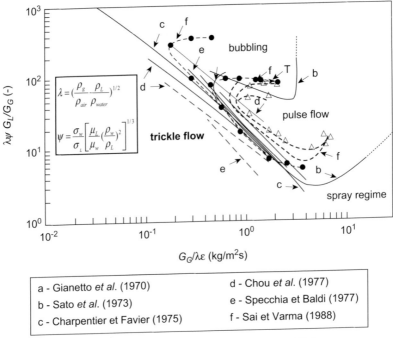

$$\lambda = (\frac{\rho_g}{\rho_{air}} \frac{\rho_L}{\rho_{water}})^{1/2}$$

$$\psi = \frac{\sigma_w}{\sigma_L}\left[\frac{\mu_L}{\mu_w}(\frac{\rho_w}{\rho_L})^2\right]^{1/3}$$

a - Gianetto et al. (1970)	d - Chou et al. (1977)
b - Sato et al. (1973)	e - Specchia et Baldi (1977)
c - Charpentier et Favier (1975)	f - Sai et Varma (1988)

FIGURE 4 Flow regime map (Saroha & Nigam, 1996).

occurs typically at $V_G < 0.75$ kg/m²s and $V_L < 12$ kg/m²s. Flow regime boundaries do not solely depend upon gas–liquid flow rates and are also a function of bed characteristics and fluid properties.

FLOW REGIME TRANSITION

In industrial practice, reactors are often operated close to the flow transition boundary between trickle and pulse flow regimes. This realizes better mass transfer rates, catalyst utilization, and enhances production capacity. It is therefore important to have adequately accurate estimation of trickle–pulse flow regime boundary. Several studies (see, for example, Attou & Ferschneider, 2000; Grosser, Carbonell, & Sundaresan, 1988; Holub, Dudukovic & Ramachandran, 1993; Wammes & Westerterp, 1990) have focused on how and when this transition occurs and its relationship with reactor hardware and operating parameters. Considering the importance of this transition boundary for design and operation of industrial trickle bed reactors, some aspects of transition such as physical picture, methods of detection, and the parameters influencing the transition boundary are briefly discussed here.

As discussed in the earlier section, pulse flow regime occurs with increase in gas or liquid flow rates. For a fixed mass flow rate of the gas, increase in liquid flow rate leads to increase in local liquid holdup to such an extent that it completely blocks the flow passage for the gas phase. The concept of flow passage blockage is discussed in several studies (Ng, 1986; Sicardi, Gerhard, & Hofmann, 1979). Such a blockage generates disturbances which propagate and grow along the length of the trickle bed. When these disturbances grow to a measurable level, pulse flow regime is observed (Krieg, Helwick, Dillon, & McCready, 1995). In other models, appearance of pulse flow is related to the instability occurring in the liquid film due to the shear exerted by the gas phase (Grosser et al., 1988; Holub et al., 1993). For fixed liquid mass flux, shear exerted by the gas phase on the liquid film drives the excess liquid in the film along the solid surface. Accumulated excess liquid generates blockage to the gas flow passage which eventually leads to pulse formation. Though measurable size of disturbance (pulse) is considered as the inception of pulse flow regime, particle-scale processes suggest that actual inception starts much earlier.

Several experimental methods have been used to detect the transition from trickle to pulse flow regime. Most of the earlier studies were based on visual observations. A change in slope of measured pressure drop or liquid holdup with respect to gas or liquid flow rates may also indicate the transition to the pulse flow regime (Bansal, Wanchoo, & Sharma, 2005; Burghardt et al., 1994; Chou, Worley, & Luss, 1977; Sicardi et al., 1979; see Fig. 5). In recent years, various online sensors and imaging techniques have been used for the identification of flow regime transition, some examples of which are conductivity probes (Tsochatzidis & Karabelas, 1994), micro-electrodes (Latifi et al., 1989), wall pressure fluctuations (Chou et al., 1977; Gunjal, Kashid, Ranade, & Chaudhuri,

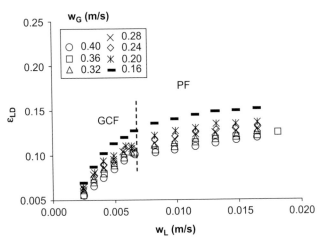

FIGURE 5 Variation of dynamic liquid holdup in trickle and pulse flow regimes (Burghardt, Bartelmus, & Szlemp, 2004). GCF: gas continuous flow; PF: pulse flow; w_G and w_L: gas–liquid superficial velocities, for the air–glycerine (30%) system. P = 1.2 MPa.

2005), computed tomography (CT), magnetic resonance imaging (MRI) (Lutran, Ng, & Delikat, 1991; Sederman et al., 2001), etc.

Experimental studies have identified several key system parameters which influence regime transition. Some of these are particle diameter and its shape factor (d_p and φ), porosity of the bed (ε) and properties of gas–liquid phases (viscosity, density, surface tension, contact angle), column diameter, and operating parameters like pressure, temperature, and gas or liquid flow rates. Inception of pulse flow regime is usually calculated based on critical liquid velocity for transition at the given gas flow rate. The fluid properties can be grouped together in the form of following dimensionless number which can be related to the flow regime transition (Baker, 1954; Bansal et al., 2005):

$$\lambda = \left[\left(\frac{\rho_G}{\rho_{air}}\right)\left(\frac{\rho_L}{\rho_W}\right)\right]^{1/2} \text{ and } \xi = \left(\frac{\sigma_W}{\sigma_L}\right)^{3.5}\left(\frac{\mu_L}{\mu_W}\right)^{0.5}\left(\frac{\rho_W}{\rho_L}\right)^{1/3} \quad (1)$$

The subscript W denotes the property of "water." Relevant correlations which may be used for estimating regime transition are summarized in Table 1. Bansal et al. (2005) have accounted for the influence of particle diameter and non-Newtonian behavior of fluids in their correlations.

Few attempts have been made to develop models and correlations based on physical understanding of the transition. Wammes & Westerterp (1990) have derived a correlation based on a criterion that assumes transition occurring at some critical dynamic holdup and it is related to the operating parameters as follows:

$$j_G^{0.26} = \frac{C}{\beta_{t,\,tr}\rho_G^{0.04}} \quad (2)$$

where, $C = 0.27$ for water–N_2 and 0.32 for 40% ethylene glycol–N_2. j_G is the gas superficial velocity in m/s and $\beta_{t,tr}$ denotes the total liquid saturation (ratio of static + dynamic liquid holdup to void volume fraction).

Another criterion is proposed by Holub et al. (1993) based on loss of the stability of laminar liquid film during the inception of the pulse flow regime. A phenomenological criterion is developed on the basis that instability of waves on the surface of the liquid film appears due to high shear exerted by the gas phase. For this purpose, Kapitza's criterion is used for calculating occurrence of surface waves and it is related to the operating parameters in the following manner (Holub et al., 1993):

$$\frac{2.90Re_L E_1^{5/11}}{\psi_L^{0.17}Ga_L^{0.41}Ka^{1/11}} \leq 1, \quad \psi_L^{0.17} = \frac{1}{\rho_L g}\frac{dP}{dZ} + 1$$
$$2.8 < Ga_L < 6.3, \quad 1 < \psi_L < 55 \quad (3)$$

TABLE 1 Empirical Correlations for Estimation of Trickle to Pulse Flow Regime Transition (λ and ξ are defined in Eq. (1))

Author	Correlations
Larachi, Laurent, Wild, and Midoux (1993) $$\Phi = \frac{1}{4.76 + 0.5\frac{\rho_G}{\rho_a}} \quad \text{and}$$ $$\psi = \frac{\sigma_W}{\sigma_L}\left(\frac{\mu_L}{\mu_W}\right)^{1/3}\left(\frac{\rho_W}{\rho_L}\right)^{2/3}$$	$$\frac{L_t\lambda\psi\Phi}{G} = \left(\frac{G}{\lambda}\right)^{-1.25}$$
Wang, Mao, and Chen (1994)	$$L_t\lambda\psi = 4.864\left(\frac{G}{\lambda}\right)^{-0.337}, \quad \frac{G}{\lambda\varepsilon} = 1$$
Dudukovic and Mills (1986)	$$L_t = \min\left[\frac{10^3 G}{\lambda\psi}5.43\frac{\varepsilon}{\psi}\left(\frac{\lambda\varepsilon}{G}\right)^{0.22}\right]$$
Bansal et al. (2005) where, $$S_1 = \frac{a_s d_p}{\varepsilon} \quad \text{and} \quad S_2 = \left(\frac{1}{d_{ps}}\right)^{\frac{1}{\phi}}$$ $$\frac{L_t\lambda\psi\varphi}{G} = \left(\frac{G}{\lambda}\right)^{-1.25}$$ $$We = \lambda_{eff}\cdot\dot\gamma_W$$ $$\dot\gamma_W = \frac{3n+1}{4n}\frac{8V_0}{\varepsilon D_e}\frac{k_i}{2}, \quad \lambda_{eff} = \frac{N_1}{2\dot\gamma\tau}$$	For Newtonian liquid phase $$\left(\frac{L_i}{G}\right)\lambda\xi\left(\frac{S_2}{S_1^2}\right)^{1/4} = 5.73\left(\frac{G}{\lambda\varepsilon}\right)^{-4/3}$$ For non-Newtonian (visco-inelastic) liquid phase $$\left(\frac{L_{t,vi}}{G}\right)\lambda\xi'\left(\frac{S_2}{S_1^2}\right)^{1/4} = 5.73\left(\frac{G}{\lambda\varepsilon}\right)^{-4/3}$$ For non-Newtonian (visco-elastic) liquid phase $$L_{t,ve} = L_{t,ve}(1 + \sqrt{2}We^{-2})$$

where the Galileo number, $Ga_L = d_p^3\rho_{L}^2 g\varepsilon^3/\mu_L^2(1-\varepsilon)^3$ and Kapitza number, $Ka = \sigma_L^3\rho_L/\mu_L^4 g$. This semi-empirical model requires knowledge of liquid phase pressure drop and Ergun constant E_1 and is suitable for estimation of the flow regime boundary for the reactors operated at or near atmospheric pressure.

Grosser et al. (1988) have used linear stability analysis of one-dimensional model (Eqs.(4) and (5)) for estimating the regime transition boundary.

$$\frac{\partial\varepsilon_i}{\partial t} + \nabla(\varepsilon_i u_i) = 0 \tag{4}$$

$$\rho_i\varepsilon_i\left(\frac{\partial u_i}{\partial t} + u_i\cdot\nabla u_i\right) = -\varepsilon_i\nabla P_i + \varepsilon_i\rho_i g + F_i \tag{5}$$

where, u is interstitial velocity of ith phase (gas or liquid).

Pressure difference in the gas and liquid phases (capillary pressure) is correlated using the Leverett's function (Scheidegger, 1974) as:

$$P_G - P_L = \left(\frac{\varepsilon}{k}\right)^{1/2} \sigma J(\varepsilon_L), \quad \text{where } J(\varepsilon_L) = 0.48 + 0.36 \ln\left(\frac{\varepsilon - \varepsilon_L}{\varepsilon_L}\right). \quad (6)$$

where, P is pressure, ε is porosity, and k is permeability of fluid. For the interphase drag force terms (F_i), the semi-empirical form of expressions provided by Saez and Carbonell (1985) were used:

$$F_G = -\left[\frac{A\mu_G(1-\varepsilon)^2\varepsilon^{1.8}}{d_p^2\varepsilon_G^{2.8}} + \frac{B\rho_G(1-\varepsilon)\varepsilon^{1.8}|u_G|}{d_p\varepsilon_G^{1.8}}\right]u_G \quad (7)$$

$$F_L = -\left\{\frac{\varepsilon - \varepsilon_s}{\varepsilon_l - \varepsilon_s}\right\}^{2.9}\left[\frac{A\mu_L(1-\varepsilon)^2\varepsilon^2}{d_p^2\varepsilon^3} + \frac{B\rho_L(1-\varepsilon)\varepsilon^3|u_L|}{d_p\varepsilon^3}\right]u_L \quad (8)$$

where, ε_{LS} is static liquid holdup, $\varepsilon_{LS} = (20 + 0.9E)^{-1}$, and $E = \rho_L g d_p^2\varepsilon^2 / \sigma(1-\varepsilon)^2$.

Though this approach uses fundamental equations of hydrodynamics, some terms appearing in the equations (capillary pressure, relative permeability, static liquid holdup, interphase drag, etc.) are based on empirical correlations. Attou and Ferschneider (2000) have developed models to reduce the empiricism associated with these parameters. Capillary pressure is expressed by them in terms of gas holdup and particle diameter as:

$$P_G - P_L = 2\sigma\left(\frac{1-\varepsilon}{1-\varepsilon_G}\right)^{1/3}\left(\frac{1}{d_p} - \frac{1}{d_{min}}\right), \quad \text{where } d_{min} = \left(\frac{\sqrt{3}}{\pi} - \frac{1}{2}\right)^{1/2} d_p \quad (9)$$

The formulation of interphase forces is based on Ergun's equation in which liquid phase is considered as an additional phase to the gas flow in the packed bed and their expressions take the following forms (Attou & Ferschneider, 2000):

Gas–liquid momentum exchange term:

$$F_{GL} = \varepsilon_G\left(\frac{E_1\mu_G(1-\varepsilon_G)^2}{d_p^2\varepsilon_G^2}\left[\frac{\varepsilon_S}{(1-\varepsilon_G)}\right]^{0.667}\right.$$
$$\left. + \frac{E_2\rho_G(U_G - U_L)(1-\varepsilon_G)}{d_p\varepsilon_G}\left[\frac{\varepsilon_S}{(1-\varepsilon_G)}\right]^{0.333}\right) \quad (10)$$

Gas—solid momentum exchange term:

$$F_{GS} = \varepsilon_G \left(\frac{E_1 \mu_G (1 - \varepsilon_G)^2}{d_p^2 \varepsilon_G^2} \left[\frac{\varepsilon_S}{(1 - \varepsilon_G)} \right]^{0.667} \right.$$

$$\left. + \frac{E_2 \rho_G U_G (1 - \varepsilon_G)}{d_p \varepsilon_G} \left[\frac{\varepsilon_S}{(1 - \varepsilon_G)} \right]^{0.333} \right) \tag{11}$$

Liquid—solid momentum exchange term:

$$F_{LS} = \varepsilon_L \left(\frac{E_1 \mu_L \varepsilon_S^2}{d_p^2 \varepsilon_L^2} + \frac{E_2 \rho_L U_G \varepsilon_S}{d_p \varepsilon_L} \right) \tag{12}$$

Linear stability analysis of one-dimensional flow equations and their sub-components (Eqs. (10)—(12)) provides prediction of the flow regime transition boundaries more accurately than empirical correlations derived from the experimental data (Attou & Ferschneider, 2000). Influence of various system parameters (bed geometry, operating parameters, and fluid properties) on the transition of regimes is briefly discussed in the following.

In smaller diameter trickle beds, inception of pulse flow regime is observed earlier than the larger diameter columns. Reactor diameter plays a significant role in the formation of blockage in the form of liquid-rich zone. It is usually difficult to operate larger diameter trickle bed reactors in a pulse flow regime. For larger diameter particles, capillary forces are less dominant than gravitational forces. Therefore the liquid holdup in the bed is substantially lower for the larger-sized particles. The transition to pulse flow therefore gets delayed for larger-sized particles (Gunjal et al., 2005). Similar phenomenon is observed in higher porosity beds (Chou et al., 1977; Sai & Varma, 1988).

Influence of gas—liquid throughput on the trickle to pulse flow regime boundary is shown in Fig. 6. At higher gas or liquid throughputs, inertial forces attenuate waves on the liquid film which are responsible for destabilization of the trickle flow pattern. Therefore, for higher gas throughputs, transition to pulse flow regime occurs at smaller liquid flow rates and vice versa (Attou & Ferschneider, 2000; Grosser et al., 1988; Ng, 1986).

Gas and liquid phase properties also have significant effect on transition boundary. Effect of capillary force can be significantly reduced by decrease in surface tension of the liquid. Experiments carried out by Charpentier and Favier (1975), Chou et al. (1977), and Wammes & Westerterp (1991) demonstrated such effects on flow regime transition boundary where water and cyclohexane were used as the liquid phases. Cyclohexane with lower surface tension (3 times lower than water) shows early inception of the pulse flow regime (see Fig. 7).

Influence of gas phase viscosity and liquid phase density on transition boundary is rather small as compared to the liquid phase viscosity and gas

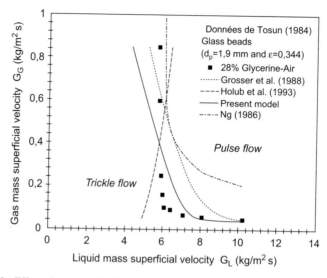

FIGURE 6 Effect of gas and liquid throughputs on trickle to pulse flow regime transition boundaries (Attou & Ferschneider, 2000).

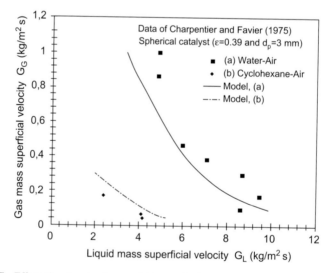

FIGURE 7 Effect of surface tension on trickle to pulse flow regime transition boundaries (Attou & Ferschneider, 2000).

phase density. Higher liquid phase viscosity leads to increase in liquid holdup, and therefore earlier inception of the pulse flow regime. Available experimental data confirm these trends (Morsi, Midoux, Laurent, & Charpentier, 1982; Sai & Varma 1988; Wammes & Westerterp, 1991). Most of the industrial trickle bed reactors are operated at high pressures. It is therefore important to understand the influence of operating pressure (or gas density) on the regime boundary. Al-Dahhan and Dudukovic (1994) have shown that effect of operating pressure is negligible as long as gas density is below 2.3 kg/m³. However, effect of operating pressure is significant when gas density exceeds 2.3 kg/m³. Attou and Ferschneider (2000) have attributed effect of increase in gas density to decrease in inertial forces; therefore transition boundary occurs at higher gas and liquid velocities. Experimental data of Wammes & Westerterp (1991) demonstrated the effect of operating pressure on transition boundary (see Fig. 8).

Despite years of experimental and modeling efforts, estimation of trickle to pulse flow regime transition is still not very accurate. Earlier models (Ng, 1986; Sicardi & Hofmann, 1980) assumed that inception of pulse flow occurs at pore-scale level, while recent models (Attou & Ferschneider, 2000; Grosser et al., 1988; Holub, Dudukovic, & Ramachandran, 1992) considered it as a bed-scale (macro-scale) phenomenon. Though macroscopic models have shown relatively better agreement with the experimental data, experimental data also indicate the importance of pore-scale processes in the inception of regime transition. Therefore, understanding physical phenomenon associated with inception of pulse flow is still a challenge and further efforts based on multi-scale modeling approach may lead to improvements in estimation of regime transition. In the mean time, equations and regime maps included here may be used for estimating the regimes and transition boundaries. *Estimation of key hydrodynamic parameters* is discussed in the following section.

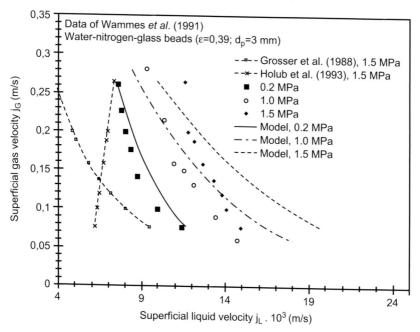

FIGURE 8   Effect of operating pressure on trickle to pulse flow regime transition boundaries.

ESTIMATION OF KEY HYDRODYNAMIC PARAMETERS

Pressure Drop

Pressure drop in trickle bed reactors is one of the most important design parameters. It is one of the key interaction indices for the overall system and therefore is often used as correlating parameter for prediction of other design parameters such as gas–liquid, liquid–solid mass transfer coefficient, wetting efficiency, and heat transfer coefficient. Two-phase pressure drop along the length of the bed is a function of (1) the reactor hardware such as column diameter, particle size and shape, and internals; (2) operating parameters such as gas–liquid flow rates (flow regime); and (3) fluid properties like density and viscosity of flowing fluid, surface tension, and surface characteristics. Operating pressure and temperature indirectly affect the pressure drop through fluid properties.

Reactor column diameter (D) has relatively lower influence on pressure drop as compared to the particle diameter (d_p). This influence is more significant for low D/d_p ratio ($D/d_p < 0.23$). For high D/d_p ratio, variation of pressure drop with column diameter is almost negligible. For low D/d_p ratio, variation of porosity near the wall plays an important role. Due to high porosity near wall, fluid bypassing occurs, resulting in a lower pressure drop. In large diameter

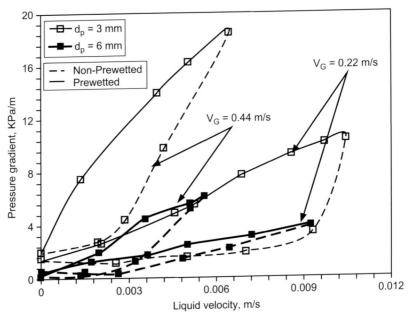

FIGURE 9 Variation of pressure drop at different bed and particle diameters for prewetted and non-prewetted conditions (Gunjal et al., 2005; $D = 11.4$ cm).

columns, uniform distribution of liquid phase is rather difficult. Liquid mal-distribution across the bed cross-section may lead to lower interaction among the phases and therefore lower pressure drop. Trickle bed reactors are often operated at low liquid flow rates which cause incomplete wetting of particles. Pressure drop for incompletely wetted particles is often less than completely wetted particles. It is highest for uniformly distributed liquid and completely wetted particles. However, at lower liquid flow rates, measured pressure drop values often show large variation due to non-uniform liquid spreading and wetting.

Particle size and shape also affect the bed pressure drop considerably. Pressure drop is less sensitive to shape factor as compared to the porosity of the bed. Denser beds (lower porosity) lead to higher pressure drop. Pressure drop increases with decrease in size of the particle. Fluid has to follow more tortuous path in the bed with smaller-sized particles. The commercial trickle beds therefore generally use particles in the range of 1 mm to 3 mm to strike app-ropriate balance of pressure drop and catalyst utilization. A sample of exper-imental results indicating pressure drop in trickle beds as a function of liquid velocity for two different particle sizes is shown in Fig. 9.

Typical variation of pressure drop with liquid flow rate at a constant gas flow rate is shown in Fig. 10. Pressure drop variation with liquid flow rate shows hysteresis behavior in trickle flow regime (see Figs. 9 and 10) which results due to different bed wetting characteristics while increasing and decreasing the flow rates. Pressure drop for non-prewetted bed is represented by the lower curve due to low interaction between gas and liquid phases. Difference between upper and

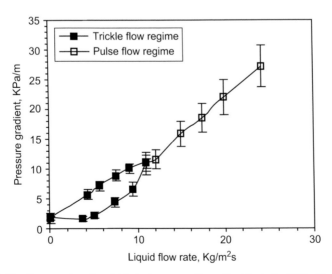

FIGURE 10   Pressure drop in trickle bed reactors (from Gunjal et al., 2005; $D = 0.141$ m, $d_p = 0.003$ m, $V_G = 0.22$ m/s).

lower branches indicates that capillary pressure is significant for non-prewetted bed which restricts the spreading of liquid over solid surface. This effect is negligible for trickle flow operation with prewetted bed or in pulse flow regime.

Reactor hardware (reactor diameter, particle size and shape, etc.) and gas–liquid throughputs have significant effect on the pressure drop. Hence most of the correlations were expressed in terms of Reynolds number of gas and liquid phases. For example, correlation of Kan and Greenfield (1979) for pressure drop is

$$
\frac{\left(\dfrac{\Delta P}{L}\right)_{LG}}{\left(\dfrac{\Delta P}{L}\right)_{G}} = 0.024 d_p \mu_L \left(\frac{\varepsilon_B}{1-\varepsilon_B}\right)^3 \left(\frac{Re_G We_G}{Re_L}\right)^{-1/3}
\tag{13}
$$

where ε_B is the bed porosity, Re, the Reynolds number based on particle $Re_\alpha = (\rho_\alpha U_\alpha d_p / \mu_\alpha)$, and We, the Weber number defined as $We_G = (\rho_G U_G^2 d_p / \sigma_G)$. The subscripts L and G denote liquid and gas phases, respectively. The above correlation is valid for trickle as well as pulse flow regimes operated at atmospheric pressure and temperature conditions. Several correlations are available for wider range of properties of gas, liquid, and solids (see, for example, those listed in Saez & Carbonell, 1985; Sai & Varma, 1987). Note that these correlations require knowledge of single-phase pressure drop which itself is a function of bed properties. Another class of correlations uses modified Lockhart–Martinelli number instead of single-phase pressure drop. The modified Lockhart–Martinelli number is defined as:

$$
X_L = \frac{1}{X_G} = \frac{U_L}{U_G}\left(\sqrt{\frac{\rho_L}{\rho_G}}\right)
\tag{14}
$$

Using this number, Ellman, Midoux, Laurent, and Charpentier (1988) have proposed the following correlation which is applicable for low as well as high interaction flow regimes for operating pressure in the range of 0.1–10 MPa:

$$
\frac{\Delta P}{L} = \frac{2\rho_G U_G^2 [A(X_G \xi_1)^j + B(X_G \xi_1)^k]}{d_p},
$$

$$
\text{for low interaction regime,} \quad \xi_1 = \frac{Re_L^2}{(0.001 + Re_L^{1.5})}, \; A = 200; \; B = 85;
$$

$$
j = -1.2, \; k = -0.5,
$$

$$
\text{for high interaction regime,} \quad \xi_1 = \frac{Re_L^{0.25} We_L^{0.2}}{(1 + 3.17 Re_L^{1.65} We_L^{1.2})^{0.1}}, \; A = 6.96;
$$

$$
B = 53.27; \; j = -2; \; k = -1.5.
\tag{15}
$$

Ergun's equation is widely used for calculating single-phase pressure drop in packed beds. This has been extended for the two-phase flow through packed beds in many studies. For example, the correlation proposed by Wammes & Westerterp (1991) is given in Eq. (16) and that by Benkrid, Rode, and Midoux (1997) in Eq. (17).

$$\frac{\Delta P}{L}\frac{d_p}{(1/2)\rho_G U_G^2} = 155\frac{(1-\varepsilon)}{\varepsilon_G}\left(\frac{\rho_G U_G d_p \varepsilon}{\mu_G(1-\varepsilon)}\right)^{-0.37} \tag{16}$$

$$\frac{\Delta P}{L} = \frac{1}{\varepsilon^3}\left(\frac{\frac{U_G}{U_L}+1}{A\frac{U_G}{U_L}+1}\right)\left(\frac{E_1}{36}\left(\frac{6(1-\varepsilon)}{d_p}+\frac{4}{D}\right)^2 \mu_L U_L\right.$$

$$\left.+\frac{E_2}{6}\left(\frac{6(1-\varepsilon)}{d_p}+\frac{4}{D}\right)\rho_L U_L^2\right) \tag{17}$$

where E_1 and E_2 are parameters of Ergun equation and A is a constant which is suggested as 0.49 for the above equation (Benkrid et al., 1997).

Pressure drop in pulse flow regime exhibits significant fluctuations which can be reported in two ways: (1) Pressure drop averaged over pulse and base regions as shown in Fig. 10. The error bars (indicating fluctuations) on pressure drop data increase as pulsing increases (see Fig. 10). (2) Alternatively, separate values of pressure drop are measured for the pulse and the base regions (regions between the pulse regions) as shown in Fig. 11.

Two-phase pressure drop is naturally a function of fluid properties. Gas phase viscosity has negligible influence compared to the liquid phase

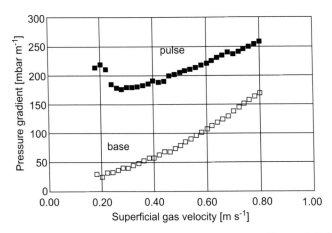

FIGURE 11 Pressure drop in pulse and base regions (Boelhouwer, Piepers, & Drinkenburg, 2002; $d_p = 6$ mm, $D = 0.11$ m, $\varepsilon = 0.38$).

viscosity. Higher viscosity of liquid leads to higher values of pressure drop. Foaming liquids are prone to produce higher gas–liquid interfacial area and therefore higher energy loss/pressure drop is associated with such systems (Larachi, Laurent, Midoux, & Wild, 1991; Midoux, Favier, & Charpentier, 1976). The following correlation proposed by Larachi et al. (1991) is valid for foaming and non-foaming liquids for 0.2–8.1 MPa operating pressure.

$$\frac{\Delta P}{L} = \frac{2G_G^2}{d_h \rho_G (X_G (Re_L We_L)^{1/4})^{3/2}} \left[31.3 + \frac{17.3}{\sqrt{X_G (Re_L We_L)^{1/4}}} \right] \tag{18}$$

where, d_h is hydraulic diameter ($d_h = d_p \sqrt[3]{16\varepsilon^3/9\pi(1-\varepsilon)^2}$) and X_G is the modified Lockhart–Martinelli number defined in Eq. (14).

Influence of operating pressure on pressure drop across a trickle bed is shown in Fig. 12. Pressure drop at high pressure conditions is similar to that operated with a gas of high molecular weight under lower pressure (provided that density of gas under operating conditions is similar). Figure 12 illustrates such an effect where pressure drop for the bed operated at 0.34 MPa with nitrogen gas is comparable with the bed operated at high pressure (2.13 MPa) with helium gas. Most of the previous correlations were developed based on the experimental data collected at atmospheric conditions and may not be useful for reactors operated at high pressure. The correlations proposed by Holub et al. (1992) (Eq. (19)), Larachi et al. (1991) (Eq. (18)), and Wammes & Westerterp (1991) (Eq. (16)) are applicable over a range of operating pressures (0.1–10 MPa).

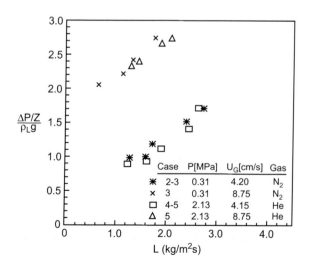

FIGURE 12 Effect of operating pressure on pressure drop in trickle bed reactors (Al-Dahhan & Dudukovic, 1994). System: hexane, nitrogen, and helium, with $D = 0.022$ m; $d_p = 1.14$ mm glass beads.

Case	P[MPa]	U_G[cm/s]	Gas
＊ 2-3	0.31	4.20	N_2
× 3	0.31	8.75	N_2
□ 4-5	2.13	4.15	He
△ 5	2.13	8.75	He

Correlation proposed by Holub et al. (1992) is as follows:

$$\psi_L = \frac{\Delta P}{\rho_L g Z} + 1 = \left(\frac{\varepsilon}{\varepsilon_L}\right)^3 \left[\frac{E_1 Re_L}{Ga_L} + \frac{E_2 Re_L^2}{Ga_L}\right],$$

$$\psi_G = \frac{\Delta P}{\rho_G g Z} + 1 = \left(\frac{\varepsilon}{\varepsilon - \varepsilon_L}\right)^3 \left[\frac{E_1 Re_G}{Ga_G} + \frac{E_2 Re_G^2}{Ga_G}\right],$$

$$\psi_L = 1 + \frac{\rho_G}{\rho_L}[\psi_G - 1]. \tag{19}$$

Prediction of pressure drop in trickle bed reactors is an important design parameter. It is sensitive to the particle packing characteristics as well as to the properties of the flowing fluids. Many earlier pressure drop correlations for gas–liquid flow through packed beds were based on the data obtained with packing like Raschig rings and hollow cylinders. These correlations often under-predict the pressure drop for particles with spherical or trilobe shapes. It is therefore important to select appropriate pressure drop correlation based on prevailing flow regime, reactor scale, particle sizes and shapes, properties of the fluid, and operating pressure and temperature.

Liquid Holdup

Liquid holdup in trickle bed reactor is expressed in two ways : (i) total liquid holdup (ε_L) defined as volume of liquid per unit bed volume and (ii) liquid saturation (β_L) defined as volume of liquid per unit void volume instead of unit bed volume. Liquid holdup is composed of two parts: dynamic liquid holdup (ε_{Ld}) and static liquid holdup (ε_{Ls}). Static liquid holdup is the volume of liquid per unit volume of bed which remains in the bed after draining the bed. Many other design parameters of trickle bed reactors, like wetting efficiency and heat and mass transfer coefficients, are dependent on liquid holdup. The prevailing liquid holdup in the bed also controls the liquid phase residence time and therefore conversion of the reactants. It is therefore essential to understand how liquid holdup varies with (1) reactor hardware such as column diameter, particle size and shape, and internals, (2) operating parameters such as gas and liquid flow rates, and (3) physico-chemical properties of fluids. Available information on this is briefly discussed in the following.

Liquid holdup is sensitive to variation in bed diameter at low D/d_p ratio and increases with the bed diameter for a given particle size (Gunjal et al., 2005). For smaller column diameters, flow bypassing results in lower pressure drop which causes low gas–liquid interaction. Therefore, liquid holdup is higher for the lower diameter columns where flow bypassing occurs. However, for larger diameter columns, liquid holdup is less sensitive to the column diameter due to negligible wall effects. Liquid holdup is considerably sensitive to particle diameter than bed diameter. Specific area of solid particles is higher for the smaller-sized particles (Gunjal et al., 2005; Rao, Ananth, & Varma, 1983); leading to higher liquid phase retention and holdup. Following correlations proposed by Burghardt, Bartelmus, Jaroszynski, and Kolodziej (1995) (Eq. (22)) and Specchia and Baldi (1977) (Eqs. (20) and (21)) demonstrate the dependency of dynamic liquid holdup to the particle diameter and interfacial area in low and high interaction regimes. These correlations are valid for a bed operated at atmospheric conditions.

For low interaction regime:

$$\beta_d = 3.86 Re_L^{0.545} (Ga^*)^{-0.42} \left(\frac{a_s d_p}{\varepsilon}\right)^{0.65}, \quad 3 < Re_L < 470 \qquad (20)$$

where β_d is dynamic liquid holdup, a_s is specific surface area of solid, and $Ga^* = d_p^3 \rho_L (\rho_L g + \Delta P / \Delta Z / \mu_L^2)$.

For high interaction regime:

$$\beta_d = 0.125 \left(\frac{Z}{\psi^{1.1}}\right)^{-0.312} \left(\frac{a_s d_p}{\varepsilon}\right)^{0.65}, \quad 1 < Z/\psi^{1.1} < 500$$

$$\psi = (\delta_W/\delta_L) \left[\left(\mu_L/\mu_W\right)(\rho_W/\rho_L)^2\right]^{0.33} \qquad (21)$$

where δ is pressure drop per unit length for a single-phase flow.

$$\beta_d = 1.125 (Re_G + 2.28)^{-0.1} (Ga_L')^{-0.5}$$

$$\times \left(\frac{a_s d_p}{\varepsilon}\right)^{0.3} \tan h\left(48.9(Ga_L')^{-1.16} Re_L^{0.41}\right),$$

$$\text{for } 2 < Re_L < 62, 0 < Re_G < 103, 51 < Ga_L' < 113 \qquad (22)$$

where, $Ga' = d_p/(\mu_L^2/(g\rho_L^2))^{1/3}$.

Similar to pressure drop, liquid holdup also exhibits hysteresis behavior with variation in gas and liquid flow rates. The magnitude of hysteresis depends on the extent of prewetting of the bed. Extent of hysteresis is highest when the bed is at fully dry conditions.

Gas and liquid phase throughputs have significant effect on the liquid holdup. In the case of low interaction regime (both gas and liquid flow rates are lower), liquid holdup is insensitive to the gas flow rate. At moderate and high gas and liquid flow rates, liquid holdup decreases with increase in gas flow rate. Contrary to this trend, the liquid holdup increases with liquid flow rate because of displacement of gas phase by the liquid. In a trickle flow regime, this displacement occurs till liquid occupies maximum possible region. Therefore, rate of increment in holdup with liquid flow rate is higher compared to the pulse flow regime (see Fig. 13). In pulse flow regime, marginal increment in holdup

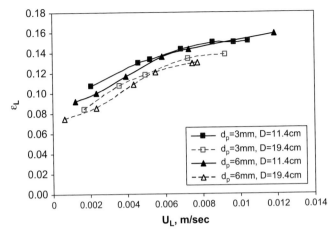

FIGURE 13   Effect of gas and liquid flow rates on dynamic liquid holdup (Burghardt et al., 2004).

values with liquid flow rate indicates that a scope for gas phase replacement is limited compared to the trickle flow regime. Thus, the decrease in liquid holdup is attributed to increase in shear between gas and liquid phases. In the case of pulsing flow, overall liquid holdup is composed of holdup in the base region and pulse flow region. Such representation of liquid holdup for pulsing regime is illustrated in Fig. 14. In pulse flow regime, both pulse and base holdups are comparatively less sensitive to the liquid flow rate than gas flow rates. Following correlations are useful for the prediction of liquid holdup in different flow regimes for reactors operated at atmospheric conditions. Simple form of

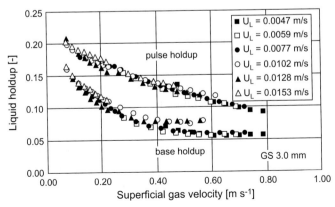

FIGURE 14   Effect of gas and liquid flow rates on dynamic liquid holdup (Boelhouwer et al., 2002). System: $D = 0.05$ m; $d_p = 3$ mm sphere, $\varepsilon = 0.4$.

correlation proposed by Wammes & Westerterp (1991) is based on Reynolds and Galileo number as follows:

$$\beta_{nc} = 16.3Re_L^c Ga_L^d$$
$$c = 0.36 \text{ and } d = -0.39 \quad \text{for } Re < 11$$
$$c = 0.55 \text{ and } d = -0.42 \quad \text{for } Re > 15 \tag{23}$$

where, β_{nc} is non-capillary liquid saturation.

Similar to the pressure drop, liquid holdup can also be related with the Lockhart–Martinelli number (X) as shown in the correlation proposed by Morsi et al. (1982). This correlation derived from the experimental data collected using air and various liquids (cyclohexane, kerosene, and poly-ethyleneglycol) for D/d_p=20.83 is given here:

$$\beta_t = 0.66X^{0.81}/(1 + 0.66X^{0.81}), \quad 0.1 < X < 80$$
$$\beta_t = 0.92X^{0.3}/(1 + 0.92X^{0.3}), \quad 0.05 < X < 100$$
$$\beta_t = 4.83X^{0.58}/(1 + 4.83X^{0.58}) \tag{24}$$

where β_t is the total liquid saturation and X is the Lockhart–Martinelli number ($X = (dP/dz)_L/(dP/dz)_G$). Rao et al. (1983) have extended this approach to include the effect of various shapes of the particles (spherical, cylindrical, and Raschig ring) for the low and high interaction regimes. They arrived at the following generalized form of correlation:

$$\beta_t = cX^b a_s^{1/3}$$
$$c = 0.4, \ b = 0.23 \quad \text{for trickle flow regime}$$
$$c = 0.4, \ b = 0.27 \quad \text{for pulse flow regime}$$
$$c = 0.38, \ b = 0.2 \quad \text{for bubbly flow regime} \tag{25}$$

where a_s is specific surface area of solid.

Liquid holdup is sensitive to the density and viscosity of the gas and liquid phases with varying extent. For example, holdup marginally changes with increase or decrease in liquid density or gas viscosity. However, it is considerably sensitive to the viscosity of the liquid phase and density of the gas phase. Liquid holdup is higher for viscous liquids and in such cases liquid–solid shear plays a greater role than the gas–liquid interactions. Most of the proposed correlations were expressed in terms of Reynolds number, which account for variation in the viscosity of the liquid or gas phase. Another important parameter that affects the liquid phase holdup is surface tension and foaming properties of the liquid phase. For liquid with surfactants, lowering the liquid surface tension enhances the spreading factor and liquid holdup decreases due to higher gas–liquid interaction. However, surface tension effects are considerable only for smaller-sized particles (1–2 mm) where capillary forces are stronger in the small interstitial spaces. Liquid holdup can be related with surface tension property of liquid by means of Weber number. Correlation proposed by Clements (1978) and Clements and Schmidt (1980) has shown such dependency for low and high interaction regimes as:

$$\varepsilon\beta_d = 0.111(We_G Re_G / Re_L)^{-0.034} \tag{26}$$

$$\beta_d = 0.245(We_G Re_G / Re_L)^{-0.034} \tag{27}$$

Liquid holdup is also considerably sensitive to the operating pressure or gas density. Gas—liquid shear is higher for bed operating at elevated pressure which causes a decrease in liquid holdup. However, this effect is negligible at lower gas velocities ($V_G < 0.01-0.02$ kg/m^{2}s) compared to the higher gas flow rates. At high pressure, liquid holdup decreases with increase in reactor operating pressure. Correlations proposed by Ellman, Midoux, Wild, Laurent, and Charpentier (1990) (Eq. (29)), Holub et al. (1992, 1993) (Eq. (19)), and Larachi et al. (1991) (Eq. (28)) are suitable for predicting liquid holdup at elevated pressures:

Larachi et al. (1991)

$$\log(1 - \beta_e) = -\frac{1.22 We_L^{0.15}}{Re_L^{0.20} X_G^{0.15}} \tag{28}$$

where, β_e is the external liquid saturation.

Ellman et al. (1990)

$$\log(\beta_{nc}) = -R X_G^m Re_L^n We_L^p \left(\frac{a_v d_k}{1 - \varepsilon}\right)^q,$$

For high interaction regime, $R = 0.16$, $m = 0.325$, $n = 0.163$, $p = -0.13$, and $q = -0.163$

For low interaction regime, $R = 0.42$, $m = 0.24$, $n = 0.14$, $p = 0$, and $q = -0.14$ (29)

a_v is the bed specific interfacial area and $d_k = d_p \sqrt[3]{16\varepsilon^3/9\pi(1 - \varepsilon)^2}$.

Predicted liquid holdup values using above correlations are compared in Fig. 15. Correlation proposed by Ellman et al. (1990) predicts significantly higher holdup compared to the predictions using correlations by Holub et al. (1992), Larachi et al. (1991), and Wammes & Westerterp (1991). Accurate estimation of the liquid holdup is desired because it is one of the key parameters in the estimation of mean residence time and conversion in the reactor.

Wetting of the Catalyst Particles

In trickle bed reactors, extent of wetting of the catalyst particles (wetting efficiency) is another important parameter required for design calculations.

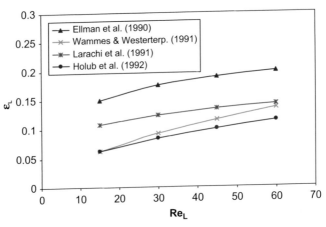

FIGURE 15 Comparison of predicted liquid holdup using various correlations for $U_G = 0.2$ m/s.

Among the various types of multiphase reactors, this phenomenon is unique in the trickle bed reactors and its quantification is a rather difficult task. Liquid flows non-uniformly over the catalyst particles leading to different degrees of wetting. Two types of wetting phenomenon are generally observed in the trickle bed reactors: external wetting of the catalyst particles and internal wetting of the particles. External wetting of the particles is a measure of fraction of catalyst surface covered by the liquid film, while internal wetting is the fraction of internal catalyst surface area covered by the liquid phase. In a trickle flow regime, external wetting is indispensable while in many situations internal wetting is not complete in spite of capillarity effects. Schematic of wetting behavior in trickle bed reactors is shown in Fig. 16. Three different wetting behaviors are observed viz., complete wetting, partial wetting, and incomplete internal wetting of particles.

External wetting can also affect some of the hydrodynamic properties in various ways and hence performance of the reactors. Presence of liquid film (like in trickle flow regime) over the catalyst surface restricts access of the gas phase reactants to the active sites. Partial wetting conditions have significant impact on the performance of the reactors. If the limiting reactant(s) is present in the liquid phase then reaction rates are directly proportional to the extent of wetting of the bed and hence partial wetting has negative effect on the performance of the reactor. If the limiting reactants are present in the gas phase, reaction rates get enhanced due to direct contact of gas phase reactants with active sites over the unwetted surface, since the catalyst particle is completely wetted internally due to capillary effects. In this case, particle wetting gives positive effect on the reactor performance. For volatile reactants, gas phase reaction can also be significant and in such a situation partial wetting was found to be beneficial. However, reaction rate analysis for such systems is fairly

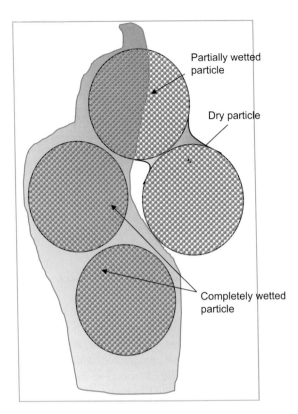

FIGURE 16 Wetting behavior of particles.

Partially wetted particle

Dry particle

Completely wetted particle

complex and requires accurate estimation of the wetted fraction. Operating parameters affecting the wetting and correlations for its accurate estimation are provided in the following.

Wetting efficiency of trickle bed reactors is defined as percent wetting of catalyst external surface area (Colombo, Baldi, & Sicardi, 1976; Herskowitz, Carbonell, & Smith, 1979; Schwartz, Weger, & Dudukovic, 1976). Measurement of wetting can be broadly categorized as direct measurement method and indirect measurement method. Photographic technique, dye injection method, computed tomography (CT) method, and magnetic resonance imaging (Sederman and Gladden, 2001) are some of the commonly used techniques for wetting efficiency. However, there are some indirect measurement techniques which are cost-effective as well as applicable for larger industrial units (for example, tracer techniques or methods based on reaction or mass transfer studies; by Colombo et al., 1976; Llano, Rosal, Sastre, & Díez, 1997, etc.). Although wetting phenomenon in trickle bed reactors is primarily governed by liquid throughputs (see Fig. 17), reactor hardware and operating parameters affect the wetting efficiency of trickle bed considerably and the effect of such parameters is discussed below.

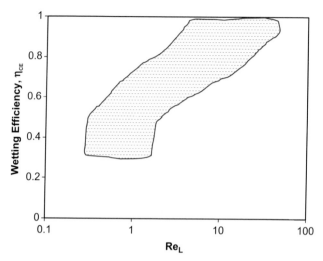

FIGURE 17 Typical variation of wetting efficiency with liquid phase Reynolds number (Herskowitz and Smith, 1983; Satterfield, 1975).

Reactor hardware, such as bed and particle diameter and porosity of bed, have significant effect on wetting efficiency of the bed. Apart from these parameters, liquid distributor is another parameter which affects wetting behavior indirectly. Non-uniform liquid distribution due to poor designing of distributor leads to liquid maldistribution. Some of the industrial-scale reactors are large in diameters (2—6 m diameter). For such large reactors, even a small variation in vertical orientation during installation of reactors leads to large-scale liquid maldistribution. Liquid maldistribution is directly related with wetting at least in a certain zone of the bed near the inlet; however, quantitative information on maldistribution is not readily available.

Apart from liquid maldistribution effects, particle diameter and porosity of the bed affect the wetting efficiency of the bed. Wetting efficiency decreases with increase in particle diameter. This trend can be attributed to two parameters: capillary pressure and liquid holdup. For smaller-sized particles, larger solid—liquid interaction leads to better spreading. Several studies reported this trend (Baussaron, Julcour-Lebigue, Wilhelm, Delmas, & Boyer, 2007; Dudukovic & Mills, 1986; El-Hisnawi, 1981). For trickle bed reactors where better wetting is critical, use of smaller-sized particles can improve wetting considerably at the expense of increase in pressure drop. Wu, Khadilkar, Al-Dahhan, and Dudukovic (1996) have suggested the use of inert fine particles along with catalyst particles to improve wetting efficiency without much increase in pressure drop. Effect of particle shape on wetting efficiency is negligible compared to the particle size (Al-Dahhan & Dudukovic, 1995; Baussaron et al., 2007). Correlations proposed by Burghardt et al. (1995) and

Mills and Duduković (1981) show the dependency of wetting efficiency on particle diameter.

$$\eta_{CE} = 1.0 - \exp\left(-1.35Re_L^{0.333}Fr_L^{0.235}We_L^{-0.17}\left(\frac{a_sd_p}{(1-\varepsilon_s)^2}\right)^{-0.0425}\right) \quad (30)$$

Or in terms of operating variables as:

$$\eta_{CE} = 0.038^*L^{0.22}G^{-0.083}d_p^{-0.373} \quad (31)$$

where, η_{CE} is external contacting efficiency, Fr is Froude number ($Fr = U^2/gd_p$), and ε_s is the solid holdup.

Besides particle diameter, external wetting efficiency is considerably sensitive to the gas—liquid flow rates. Most of the trickle bed reactors are operated at very low liquid mass flow rate (0.01—2 kg/m^{2}s). At such low liquid flow rates, liquid is insufficient to cover the catalyst surface and hence partial wetting is inevitable under such conditions. Effect of liquid flow rate is significant on wetting efficiency of the bed. Detailed analysis carried out by Sederman and Gladden (2001) shows that wetting efficiency initially changes at higher rate with liquid flow rate mainly because of the formation of new rivulets. It, however, reaches a saturation point where instead of formation of new rivulets, size of existing rivulets increases which improves wetting. Therefore, relative rate of increase of wetting efficiency is low at higher liquid throughputs. Dependency of wetting efficiency on liquid flow rate is represented by following correlations given by Alicilar et al. (1994), A.A. El-Hisnawi, Dudukovic, and Mills (1982), and A.E. El-Hisnawi, Dudukovic, and Mills (1982).

$$\eta_{CE} = 1.617Re_L^{0.146}Ga_L^{-0.0711} \quad (32)$$

$$\eta_{CE} = 1 - \frac{25.6}{Re_L^{0.56}} \quad (33)$$

Not many studies have reported the effect of gas flow rate on wetting efficiency. Gas velocity shows different trends for variation in wetting efficiency. Two changes occur with increase in gas velocity viz., wetting efficiency decreases due to decrease in liquid holdup and wetting efficiency improves due to increase in gas—liquid shear. Negative effect of gas velocity is reported by Baussaron et al. (2007), Burghardt et al. (1995), Burghardt, Kolodziej, and Jaroszynski (1990), and Sederman and Gladden (2001) and they proposed the following correlation:

$$\eta_{CE} = 3.38^*Re_L^{0.22}Re_G^{-0.83}\left(d_p\sqrt{\frac{g\rho_L^2}{\mu_L^2}}\right)^{-0.512} \quad (34)$$

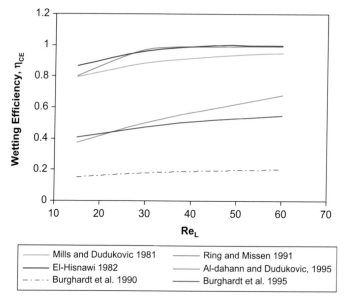

FIGURE 18 Wetting efficiency of particles predicted using various correlations.

Correlation proposed by Al-Dahhan and Dudukovic (1994) suggests an indirect effect of gas flow rate on wetting efficiency (Eq. 35). Their correlation for wetting efficiency is based on two-phase pressure drop which increases with gas flow rates and the experimental data collected at elevated operating pressure where one would expect higher gas–liquid shear:

$$\eta_{CE} = 1.104 Re_L^{0.33} \left[\frac{1 + [\Delta P/Z]/\rho_L g}{Ga_L} \right]^{1/9} \tag{35}$$

Very few studies have addressed the effect of gas density/operating pressure on wetting efficiency of the particles (Al-Dahhan & Dudukovic, 1995; Ring & Missen, 1991). Catalyst wetting efficiency improves with increase in operating pressure (higher density of gas leads to enhanced gas–liquid shear and therefore enhanced spreading of liquid).

Values of wetting efficiency calculated using various correlations are compared with each other in Fig. 18 (for trickle bed operated at atmospheric conditions and $Re_G = 86.3$). Predictions based on the correlations proposed by A.A. El-Hisnawi, Dudukovic, and Mills (1982), A.E. El-Hisnawi, Dudukovic, and Mills (1982), Mills and Duduković (1981), and Ring and Missen (1991) lie close to each other. Predictions of correlations of Al-Dahhan and Dudukovic (1995) and Burghardt et al. (1995, 1990) are much lower than those predicted by these three correlations (and lower than the expected range of >0.6). Recently Baussaron et al. (2007) and Julcour et al. (2007) measured local and average partial wetting efficiencies in trickle beds via direct measurements using image processing and PIV (particle image velocimetry). Such techniques along with MRI may provide better insight into partial wetting.

Gas–Liquid Mass Transfer Coefficient

Trickle flow operation is a rather "silent" class of operation as there is no rigorous mixing mechanism present like other multiphase catalytic systems as stirred tank, slurry bubble column reactors, or even like the high interaction flow regime. Therefore, mass transfer rates are lower than other reactors and often become rate-limiting in the performance of the trickle bed reactors. Many of the industrial reactions carried out in trickle bed reactors are reported to be falling under mass transfer-limiting conditions (for example, hydrogenation of α-methyl styrene and sulfur dioxide oxidation). Three types of mass transfer rates are relevant for trickle bed reactors, i.e., gas–liquid, liquid–solid, and gas–solid mass transfer rates. In gas–liquid mass transfer process, the liquid side mass transfer rate is often rate-limiting. Gas–liquid mass transfer rates are dependent on various system parameters such as particle diameter, gas–liquid flow rates, properties of fluids, and the reactor operating conditions. Based on these parameters several studies have reported correlations for gas–liquid mass transfer coefficients (see, for example, Cassanello, Larachi, Laurent, Wild, & Midoux, 1996; Charpentier & Favier, 1975; Fukushima & Kusaka, 1977b; Gianetto, Specchia, & Baldi, 1973; Larachi, Cassanello, & Laurent, 1998; Midoux, Morsi, Purwasasmita, Laurent, & Charpentier, 1984; Wammes & Westerterp, 1991; Wild, Larachi, & Charpentier, 1992). Key aspects are discussed in the following.

Reactor configuration (diameter, height) has rather limited role in altering gas–liquid mass transfer rates compared to the influence of particle size. The gas–liquid mass transfer rate increases with decrease in particle size. Two-phase pressure is an indication of extent of interaction among the flowing phases. For smaller-sized particles, this interaction is considerably higher and therefore the gas–liquid mass transfer rates. Some of the correlations of mass transfer include pressure drop (see, for example, a correlation proposed by Iliuta and Thyrion, 1997 which is reproduced here as Eq. (36)).

$$k_{GL}a_{GL} = 0.0036\left(\frac{U_L}{\varepsilon_{L,d}}\frac{\Delta P}{Z}\right)^{0.35} \tag{36}$$

where $\varepsilon_{L,d}$ is the dynamic liquid holdup.

Instead of expressing mass transfer correlation in terms of unknown pressure drop (which may have to be estimated using other correlations), Lange (1978) proposed a correlation based on known parameters as:

$$k_{GL}a_{GL} = 0.33D_{H_2,L}\left(\frac{d_R}{d_p}\right)^{0.46}\left(\frac{\rho_L u_L}{\mu_L}\right)^{0.14}Sc_{H_2L}^{0.5} \tag{37}$$

where D_{H_2L} is the diffusivity for hydrogen and $Sc_{H_2,L}$ is Schmidt number $(\mu_L \rho D_{H_2L})$.

Apart from particle size, gas–liquid mass transfer coefficient is also considerably sensitive to gas–liquid throughputs. Gas–liquid flow rate enhances the interaction between gas and liquid phases and spreading of liquid. This leads to an increase in gas–liquid interfacial area. Fig. 19 shows the variation of calculated mass transfer rates (based on the correlation proposed by Wild et al., 1992) with gas and liquid flow rates. The correlation proposed by Wild et al. (1992) is applicable for trickle, pulse, and transition flow regimes and is given below as Eqs. (38)–(40).

Low interaction regime:

$$\frac{k_{GL}a_{GL}d_k^2}{D_{AL}} = 2 \times 10^{-4}\left(X_G^{1/4} Re_L^{1/5} We_L^{1/5} Sc_L^{1/2}\left(\frac{a_v d_k}{1-\varepsilon}\right)^{1/4}\right)^{3.4} \tag{38}$$

High interaction regime:

$$\frac{k_{GL}a_{GL}d_k^2}{D_{AL}} = 0.45\left(X_G^{1/2} Re_L^{4/5} We_L^{1/5} Sc_L^{1/2}\left(\frac{a_v d_k}{1-\varepsilon}\right)^{1/4}\right)^{1.3} \tag{39}$$

Transition regime:

$$\frac{k_{GL}a_{GL}d_k^2}{D_{AL}} = 0.091\left(X_G^{1/4} Re_L^{1/5} We_L^{1/5} Sc_L^{3/10}\left(\frac{a_v d_k}{1-\varepsilon}\right)^{1/4}\right)^{3.8} \tag{40}$$

Gas–liquid mass transfer rates are also dependent on properties of the gas–liquid phases such as liquid phase surface tension and viscosity, diffusion

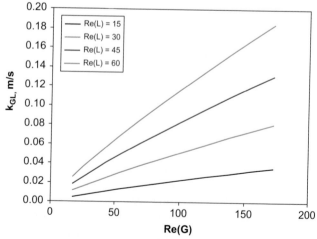

FIGURE 19   Typical variation of gas–liquid mass transfer coefficient with liquid phase Reynolds number (calculated based on Wild's correlation Eq. (38)).

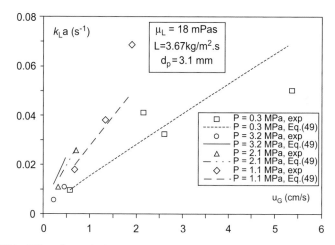

FIGURE 20   Effect of gas velocity and operating pressure on gas–liquid mass transfer coefficient (Larachi et al., 1998). System: ethylene triglycol + diethylamine 0.05 kmol, nitrogen gas with polypropylene extrudates of $d_p = 3.37$ mm [Eq.(49) in the legend refers to the equation number reported by Larachi et al. 1998].

coefficient, and gas phase density. Mass transfer rates are not very sensitive to parameters like gas phase viscosity and liquid density. Mass transfer rates vary considerably for foaming and non-foaming liquids. Foaming liquid leads to higher interfacial area and therefore higher mass transfer rates (Midoux et al., 1984; Morsi et al., 1982). For the case with foaming liquids, higher operating pressure suppresses foaming and leads to lower mass transfer rates. However, for the non-foaming liquids, operating pressure has relatively less influence on mass transfer rates. Increase in operating pressure or gas density enhances the mass transfer rates and this can be attributed to increase in gas holdup and interfacial area at high pressure (Larachi et al., 1998). At high operating pressure and moderate gas flow rate, gas phase penetrates into the liquid film in the form of tiny bubbles. This process increases the gas holdup as well as gas–liquid interfacial area. In addition to this, liquid spreading is higher at elevated operating pressure (Al-Dahhan & Dudukovic, 1994) which improves wetting as well as gas–liquid interfacial area. Experimental data collected by Larachi et al. (1998) demonstrate the effect of operating pressure on gas–liquid mass transfer rates; shown here as Fig. 20.

Liquid–Solid Mass Transfer Coefficient

Liquid–solid mass transfer is important for many trickle bed reactors and has been extensively studied (see, for example, Chou, Worley, & Luss, 1979; Kawase & Ulbrecht, 1981; Lakota & Levec, 1990; Latifi, Laurent, & Storck, 1988; Specchia, Baldi, & Gianetto, 1978). In trickle flow operations, liquid–solid mass transfer rates are dependent on the extent of liquid contact with available solid surface. With increase in liquid flow rate, wetting behavior improves and therefore the liquid–solid mass transfer rates. Given the fact that wetting factor cannot be separated from mass transfer rates, several correlations, therefore, include the wetting factor as a parameter along with the liquid–solid mass transfer coefficient (see, for example, Chou et al., 1979; Latifi, Naderifar, Midoux, & Le Mehaute, 1994; Rao & Drinkenburg, 1985). Usually the liquid–solid mass transfer coefficient is expressed in terms of Schmidt number ($Sc = \mu/\rho D$) and Reynolds number ($Re = \rho U d_p/\mu$) with or without wetting factor as follows:

$$\eta_{CE} Sh = a(Re)^b Sc^{1/3} \qquad (41)$$

where Sh is Sherwood number ($Sh = k_{LS} d_p/D$). Reported values of parameters appearing in this equation are listed in Table 2.

TABLE 2 Constants Proposed by Various Studies for Eq. (41) with Correction Factors

Author	η_{CE}	a	b	Other Factors
Chou et al. (1979)	0–1	0.43 0.72	0.22 0.54	$Re_L = Ko$ (high interaction regime) Extra term $Re_G^{0.16}$ (trickle flow regime)
Speechia et al. (1976)	1	2.79	0.7	$Re_L = Re_L/a_s$
Hirose et al. (1976)	1	8 0.53	0.5 58	($Re_L < 200$) ($Re_L > 200$)
Satterfield et al. (1978)	0–1	0.815	0.822	$Sh = Sh/(1-\varepsilon)$
Kawase and Ulbrecht (1981)	1	0.6875	0.33	—
Delaunay, Storck, Laurent, and Charpentier (1982)	0–1	1.84	0.48	—
Tan and Smith (1982)	1	4.25	0.48	—
Rao and Drinkenburg (1985)	0–1	0.24	0.75	$Re_L = Re_L/\varepsilon_L$
Latifi et al. (1988)	1	1.2	0.533	$Sh = Sh[\varepsilon_B/(1-\varepsilon_B)]$
Lakota and Levec (1990)	1	0.847	0.495	$Re_L = Re_L(\varepsilon_B/h_d)/(1-\varepsilon_B)$

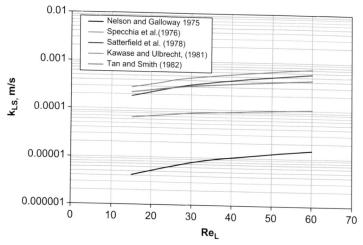

FIGURE 21 Comparison of prediction of liquid–solid mass transfer coefficient using correlations proposed by various studies.

Nelson and Galloway (1975) reported the effect of reactor and particle diameter on liquid–solid mass transfer coefficient as:

$$k_{LS}a_S\varepsilon_S = 0.75D_{H_2,L}\left(\frac{d_R}{d_p}\right)^{0.82}\left(\frac{\rho_L u_L}{\mu_L}\right)^{0.9} Sc_{H_2L}^{0.5} \tag{42}$$

Liquid–solid mass transfer rate is quite sensitive to particle diameter and increases with decrease in particle size (Goto & Smith, 1975; Hirose, Mori, & Sato, 1976; Specchia et al., 1978). Liquid–solid mass transfer rates are considerably sensitive to the liquid mass flow rates compared to the gas mass flow rates. Mass transfer rate increases with liquid mass flow rates and significant enhancement is observed at lower liquid flow rates than higher. Highfill and Al-Dahhan (2001) have shown that mass transfer rates are higher at the central portion of the column where liquid flow rates are supposed to be higher because of segregation. Gas phase has negligible effect on liquid–solid mass transfer rates and Specchia et al. (1978) have reported that these effects are considerable only at low gas flow rates. Comparison of predicted values of liquid–solid mass transfer coefficients by different correlations as a function of liquid phase Reynolds number is shown in Fig. 21.

Gas–Solid Mass Transfer

Gas–solid mass transfer coefficient is required in modeling of trickle bed reactor when catalyst particles are partially wetted. On the dry surface of the catalyst particle, there is a direct contact between gas phase reactants and the external catalyst surface. Gas–solid mass transfer has been extensively studied in single-phase reacting systems. Two key correlations are included here which may be used to estimate gas—solid mass transfer coefficients:

Nelson and Galloway (1975):

$$k_{GS} = 0.18 \left(\frac{D_{H_2, G}}{d_p}\right) \varepsilon_s^{0.33} \left(\varepsilon_s^{-0.33} - 1\right) (Re_G) \left(Sc_{H_2,G}^{2/3}\right) \qquad (43)$$

Dwivedi and Upadhyah (1977):

$$k_{GS} = 0.4548 \left(\frac{H_A u_G}{\varepsilon_B}\right) (Re_G)^{-0.4069} \left(Sc_G^{-0.667}\right) \qquad (44)$$

Axial Dispersion

Liquid distribution and mixing are other important parameters in the design of trickle bed reactors. Main factors contributing to liquid distribution and back mixing are non-uniform porosity distribution, capillarity action, dead-zones, partial wetting, and channeling or short-circuiting. These factors exhibit strong interaction among themselves and encompass wide range of spatio-temporal scales (micro-scale, meso-scale, and macro-scale). Therefore quantification of individual factors on mixing and on non-ideal flow behavior is rather a challenging task. Measurement of residence time distribution (RTD) is one of the oldest (Danckwerts, 1953) and successful methods for characterizing such flow non-idealities. Measurement of RTD is relatively straightforward compared to the processing and interpretation of RTD data. Commonly used practice is to interpret the RTD data obtained for trickle bed reactors with the axial dispersion model (ADM) which lumps all flow non-idealities into a single-fitted parameter called dispersion coefficient (D_L).

As mentioned earlier, flow non-idealities in trickle bed reactors occur due to combined effect of several macro-scale and micro-scale processes. It is possible to elicit the effect of macro-scale and micro-scale flow processes to some extent using variants of ADM models which are composed of more than one fitted parameter (2—6 parameters). Immediate extension of such a model is possible by splitting the dispersion coefficient into axial and radial directions (multi-stage dispersion model (MDM) or mixing cell model, Ramachandran & Smith, 1979). Other versions of ADM by considering random delay of fluid element due to travel in packed bed (probabilistic time model [PTB] of Buffman, Gibilaro, & Rathor, 1970) or by considering the effects of mass exchange between static and dynamic regions along with back mixing of fluid (of Van Swaaij, Charpentier, & Villermaux, 1969) have also been used. If porous particles are present in the system, dynamics of the trickle bed reactors is altered by additional lag due to associated diffusion, adsorption, and desorption processes (Iliuta & Larachi, 1999). Despite a wide variety of models, single parameter ADM is the most widely used model for characterizing system non-idealities because of its simplicity. Variation of the dispersion coefficient (in terms of Peclet number, Pe_L or Bodenstein number, Bo) with system parameters is discussed in the following.

Liquid phase Peclet number (Pe_L defined as $U_L D/D_L$) depends on key reactor parameters such as reactor diameter, length, particle diameter, and porosity. Mears (1971) has proposed a criterion to identify whether dispersion is relevant based on conversion of a reactant and the Peclet number. This criterion states that minimum reactor length required for ignoring dispersion effects as:

$$\frac{z}{d_p} > \frac{20n}{Pe_L} \ln\left(\frac{C_{in}}{C_{out}}\right) \qquad (45)$$

where, n is the reaction order, C is the concentration, and z is the reactor length. Above expression indicates that dispersion effects are considerable for higher conversions or higher order of reactions and lower Peclet numbers. Apart from reactor length, particle diameter has also significant effect on dispersion at lower liquid flow rates. Study of Fu and Tan (1996) shows that dispersion is sensitive to particle diameter at low liquid flow rates ($Re_L < 4$) and almost independent of particle diameter for higher liquid flow rates. Following correlation is proposed for low liquid flow rates,

$$Bo = 0.00014(d_h)^{-0.75}(\varepsilon)^{-1} \qquad (46)$$

where Bo is the Bodenstein number ($Bo = (Uz/D_{ax})$) and d_h is the hydraulic diameter and D_{ax} and z are dispersion coefficient and length of the column, respectively.

Pant, Saroha, & Nigam, (2000) have shown that dispersion varies considerably with shape of the particles especially with increase in liquid flow rates. For extrudates, dispersion remains unchanged for a range of Reynolds numbers ($Pe_L = 30$, for $Re_L = 24-94$). However, it changes considerably for glass beads and tablets ($Pe_L = 50-200$) for the same range of Reynolds number. Besides particle size and shape, particle porosity has also significant effect on dispersion number (Iliuta, Thyrion, & Muntean, 1996; Saroha, Nigam, Saxena, & Kapoor, 1998). Dispersion coefficient is found to be lower for porous particles than non-porous particles.

Effect of gas flow rate on dispersion coefficient is almost negligible and several studies have reported this trend (Cassanello, Martinez, & Cukierman, 1992; Pant et al., 2000; Saroha et al., 1998; Tosun, 1982). Quantitative dependency on gas flow rate is given by Hochman and Effron (1969) in the form of following correlation:

$$Pe_L = 0.034(Re_L)^{0.5}10^{0.008Re_G} \qquad (47)$$

Effect of liquid flow rate is significant on dispersion coefficient especially for lower liquid flow rates ($Re_L < 50$). Figure 22 shows the residence time

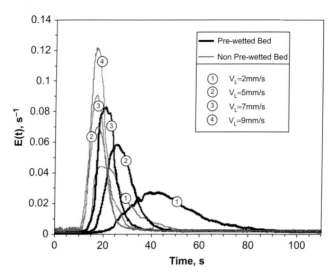

FIGURE 22 Effect of bed prewetting on dispersion behavior of trickle bed reactors (Gunjal et al., 2003). System: $D = 0.114$ m, $d_p = 3$ mm glass sphere.

distribution plots for air−water system at different liquid flow rates which clearly indicate the importance of dispersion at low liquid flow rates. This figure also demonstrates that beside liquid flow rates, bed prewetting conditions (capillary pressure) have significant effect on dispersion of liquid phase at lower liquid flow rates. Dependency of dispersion on liquid flow rates can be expressed in terms of liquid phase Reynolds number. Matsuura, Takashi, and Takashi (1976) have shown that dispersion is independent of Reynolds number for very low liquid flow rates ($Re_L < 4$) and then increases linearly with Re_L.

$$Bo = 0.43, \quad \text{for } Re_L < 150$$
$$Bo = 1.7, \quad \text{for } Re_L > 400$$
$$Bo = 0.0052Re_L - 0.332, \quad \text{for } 150 > Re_L > 400 \quad (48)$$

The generalized form of the correlation for liquid phase dispersion coefficient is

$$Bo = a(Re_L)^b(Ga_L)^c \quad (49)$$

Exponent of Reynolds number was found to be in the range of 0.3−0.6 while the exponent of Galileo number was found to be in the range of 0.2−0.7. Some of the published values of these parameters are listed in Table 3.

Dispersion effects in gas−liquid trickle flow through packed bed are significantly higher than that in single-phase flow through packed bed ($Bo = 0.5$ and 0.1 for single-phase flow and trickle flow, respectively). The

TABLE 3 Correlations for Dispersion Coefficient Proposed by Various Studies

Author	Correlations for Dispersion Number
Otake and Kunugita (1958)	$Bo = 1.9(Re_L)^{0.5}(Ga_L)^{-0.333}$
Hochman and Effron (1969)	$Pe_L = 0.042(Re_L)^{0.5}$, $4 < Re_L < 100$
Michell and Furzer (1972)	$Pe_L = (Re_L/\varepsilon)^{0.7}(Ga_L)^{-0.32}$, $50 < Re_L < 1000$
Turek, Lange, and Bush (1979)	$Bo = 0.034(Re_L)^{0.53}$
Buffham and Rathor (1978)	$Pe_L = 60(Re_L)^{0.63}(Ga_L)^{-0.73}$
Kobayashi, Kushiyama, Ida, and Wakao (1979)	$Pe_L = 60(Re_L'')^{0.63}(Ga_L'')^{-0.73}$
Cassanello et al. (1992)	$Bo = 2.3(Re_L)^{0.33}(Ga_L)^{-0.19}$
Wanchoo, Kaur, and Bansal (2007)	$Bo = 0.4548(Re_L)^{0.55}$, $4 < Re_L < 80$

dispersion effects in trickle bed reactors therefore cannot be neglected especially for low liquid flow rates, smaller length column, and higher conversions.

Heat Transfer in Trickle Bed Reactors

Most of the reactions carried out in trickle bed reactors are exothermic in nature (e.g., hydrogenation, oxidation, and hydrotreating). Removal of heat liberated due to chemical reactions from reactor becomes necessary to avoid catalyst deactivation/sintering and for safer operation. Trickle bed reactors are prone to temperature runaway conditions due to poor heat transfer rates. Therefore, operating trickle bed reactors under adiabatic conditions is one of the major challenges. Temperature uniformity in the reactor is desired in order to achieve better selectivity, product uniformity, and quality. This can be achieved by removing heat either via cooling coils or intermediate quenching, or using low boiling solvent. In trickle bed reactors, large radial temperature non-uniformities are observed and several studies have reported hotspot formation and runaway conditions (see, for example, Germain, Lefebvre, & L'Homme, 1974; Hanika, 1999; Jaffe, 1976; Weekman, 1976). Accurate estimation of heat transfer rates is therefore critical for successful design and scale-up of trickle bed reactors. Key parameters affecting heat transfer in trickle beds are discussed in the following.

Heat transfer in trickle bed reactor occurs at various levels:

- heat transfer inside the catalyst pellets where reactions occur (intraparticle heat transfer),
- heat transfer from pellet to the surrounding fluid (particle–fluid heat transfer),

- heat transfer from pellets to pellets (interparticle heat transfer), and
- heat transfer from bed to the wall of the reactor.

Considering the internal wetting and presence of liquid in the pores, in many cases, temperature is reasonably uniform inside the catalyst particle. Many studies have reported particle–fluid heat transfer coefficient for single-phase flow (see, for example, Achenbach, 1995; Briens, Del Pozo, Trudell, & Wild, 1999; Ranz, 1952, and the references cited therein). Relatively very few studies have been reported on particle–fluid heat transfer rates for trickle flow operation (see, for example, Boelhouwer, Piepers, & Drinkenburg, 2001; Marcandelli, Wild, Lamine, & Bernard, 1999). Most of the earlier studies were focused on calculating effective axial and radial heat transfer coefficients (Chu & Ng, 1985; Matsuura, Hitaka, Akehata, & Shirai, 1979; Specchia et al., 1979; Weekman & Myers, 1965) and the bed to wall heat transfer coefficients (Muroyama, Hashimoto, & Tomita, 1977; Specchia et al., 1979). Various heat transfer modes and their relationship with key parameters of trickle bed reactors are briefly discussed in the following.

Particle–fluid heat transfer coefficient is a function of gas and liquid phase velocities and thermal properties of the fluids and particles (that is, thermal conductivity of liquid, density, and viscosity of the liquid and thermal conductivity of the solids). Particle–fluid heat transfer coefficient is independent of reactor diameter. Dimensionless analysis suggests the following relationship for heat transfer coefficient and operating parameters:

$$Nu = a(Re)^b Pr^{1/3} \tag{50}$$

Boelhouwer et al. (2001) have measured heat transfer rates in 0.11 diameter column filled with 6 mm glass beads for air–water system and suggested values of $a = 0.111$ and $b = 0.8$ for trickle as well as pulse flow regimes. In pulse flow regime heat transfer rates are higher than the trickle flow regime. Detailed measurements by Boelhouwer et al. (2001) and Huang, Varma, and McCready (2004) indicate that higher liquid phase velocity is beneficial for dissipating heat from the catalyst particles in a pulse flow regime. Heat transfer rates during the passage through liquid rich zones are 3–4 times higher than passage of liquid lean zones. Figure 23 shows experimentally measured heat transfer coefficient in the liquid-rich zone of pulse flow and the base region at different gas and liquid velocities. Studies of Boelhouwer et al. (2001) also indicated that the heat transfer rates at the center of the column are lower (1.5–2.5 times) compared to the periphery of the column. This effect is attributed to the packing density and local liquid velocities.

Heat transfer was found to increase with gas as well as liquid flow rates (Boelhouwer et al., 2001; Marcandelli et al., 1999). A sample of results of Boelhouwer et al. (2001) is shown in Fig. 24. However, heat transfer rate enhancement due to increased gas flow rates is rather marginal compared to the enhancement due to increased liquid flow rates.

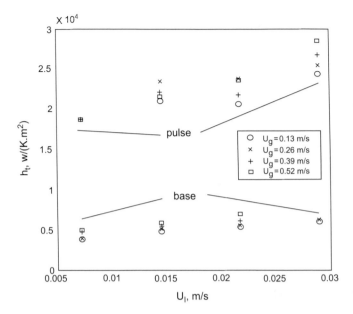

FIGURE 23 Experimental data on heat transfer coefficient measured in the liquid-rich and lean zones by Huang et al. (2004). System: $D = 7.4$ cm, $d_p = 6-8$ mm glass sphere.

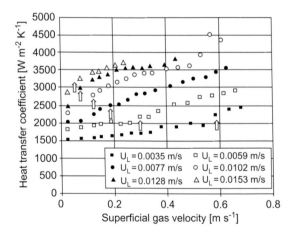

FIGURE 24 Experimental data on heat transfer coefficient at different gas and liquid velocities (Boelhouwer et al., 2001). System: $D = 0.11$ m, $d_p = 6$ mm glass sphere, $\varepsilon = 0.38$.

Effective thermal conductivity of the bed is a measure of the average thermal conductivity of the bed in the presence of gas and liquid phases in axial as well as radial directions. Most of the previous studies have focused on the effective radial thermal conductivity of the bed because of relatively higher temperature gradients in the radial direction than in the axial direction. Hashimoto, Muroyama, Fujiyoshi, and Nagata (1976) and Weekman and Myers

(1965) have related the effective radial thermal conductivity (Λ_r) to the liquid thermal conductivity and the bed properties (porosity (ε) and diameter of bed (D) and particle (d_p)) along with liquid phase Peclet (Pe_l) number and dynamic liquid holdup (β_l) in the form of following correlation:

$$\frac{\Lambda_r}{\lambda_l} = \frac{12.2}{\lambda_l} + \frac{0.000285}{\varepsilon} \frac{D}{d} \frac{Pe_l}{\beta_l} \tag{51}$$

where λ is the thermal conductivity of liquid and Pe_l is the liquid Peclet number ($Pe_l = LC_{pl}d_p/k_l$).

In addition to these parameters, Hashimoto et al. (1977) have also taken convective term into account in the form of liquid phase Reynolds number (Re_l) as follows:

$$\frac{\Lambda_r}{\lambda_l} = \frac{0.465}{\lambda_l} + \alpha_l Re_l Pr_l + 0.110 \frac{GC^*_{pg}d}{\lambda_l}$$

$$\alpha_l = \left[\frac{1}{1.9 + 0.0264(DL/\varepsilon\beta_l\mu_l)}\right] \frac{2\varepsilon}{3(1-\varepsilon)} \tag{52}$$

where C^*_{pg} is heat capacity of saturated gas.

Specchia and Baldi (1979) suggested that thermal conductivity of air and solid also needs to be accounted along with the contribution due to the gas and liquid flows. They have defined the effective radial conductivity as:

$$\Lambda_r = \lambda^o_e + \lambda_{e,g} + \lambda_{e,l} \tag{53}$$

where, λ^o_e is the bed thermal conductivity in the absence of any flow and $\lambda_{e,g}$ and $\lambda_{e,l}$ are contributions due to gas and liquid flows. With these parameters following correlation is suggested:

$$\frac{\Lambda_r}{\lambda_l} = \frac{\lambda_{SO}}{\lambda_l} + \frac{GC^*_{pg}d}{K\lambda_l} + \frac{\lambda_{rl}}{\lambda_l}, \tag{54}$$

where constants are

$$K = 8.65((1 + 19.4)(d/D)^2)$$

$$\frac{\lambda_{SO}}{\lambda_l} = \varepsilon + \frac{(1-\varepsilon)\lambda_g/\lambda_l}{0.22\varepsilon^2 + (2\lambda_g/3\lambda_S)}$$

$$\frac{\lambda_{rl}}{\lambda_l} = 24.4(\varepsilon\beta_l)^{0.87} Re_l^{0.13} Pr_l \tag{55}$$

Effective bed thermal conductivity increases with decrease in particle diameter. Effective bed conductivity depends on dynamic liquid holdup. Dynamic liquid holdup increases with liquid flow rate and decreases with gas flow rate. Data from the study of Lamine, Gerth, Legall, and Wild (1996) are shown in Fig. 25. It can be seen that the effective bed conductivity remains more or less constant in the trickle flow regime and increases as it approaches the

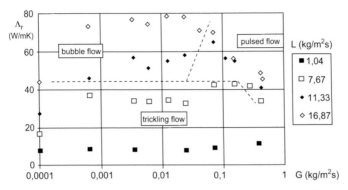

FIGURE 25 Experimental data on effective radial heat transfer coefficient at different gas and liquid velocities (Lamine et al., 1996). System: $D = 0.1$ m, $d_p = 6$ mm, air–water.

pulse flow regime. Effective thermal conductivity for the bubbly flow regime is higher than trickle flow mainly because of higher dynamic liquid holdup.

Wall to bed heat transfer rate is strongly dependent on distribution of liquid and wall flow characteristics. In trickle bed reactor, complete wetting of wall occurs occasionally and heat transfer rates are dependent upon liquid wall flow velocities. Muroyama et al. (1977) have related wall heat transfer to the liquid phase Reynolds number and Prandtl number as:

$$Nu = 0.012 Re_l^{1.7} Pr_l^{0.33} \tag{56}$$

Their study also suggests that heat transfer coefficient in the high interaction regime is a function of dynamic liquid holdup and can be predicted by the following correlation:

$$Nu = 0.092 \left(\frac{Re_l}{\varepsilon \beta_l}\right)^{0.8} Pr_l^{0.33} \tag{57}$$

Specchia et al. (1979) proposed a similar correlation except that in the high interaction regime, it remained unaffected by operating parameters as:

$$Nu = 0.057 \left(\frac{Re_l}{\varepsilon \beta_l}\right)^{0.89} Pr_l^{0.33}, \quad \text{for low interaction regime} \tag{58}$$

$$h_w = 2100 \text{ W/m}^2\text{K}, \quad \text{for high interaction regime} \tag{59}$$

It may be noted from the discussions so far that most of the key hydro-dynamic parameters interact with each other and therefore interdependent. Estimation of one hydrodynamic parameter often required estimation of other related parameters. It is therefore essential to understand implications of possible errors associated with the estimated quantities and consequences of this on subsequent usage for estimation of other related parameters. The sequence in which these different parameters are estimated may often turn out to be important. The discussion in this chapter was organized to indicate one such sequence (from flow regimes to heat transfer). However, more often than not, iterative procedure is used and multiple ways are used to cross-check and ensure consistency in the estimated values of key hydrodynamic parameters. Variety of other factors like closeness of considered system and operating conditions with those used for developing correlations, purpose of estimation, error bars associated with the data used while developing correlations, and prior experience determine the selection of particular set of correlations and methodology for estimating the required set of hydro-dynamic parameters.

SUMMARY

Accurate estimation of hydrodynamic parameters is an indispensable step for the reactor design, performance evaluation, and scale-up studies. Hydrody-namics of trickle bed reactors is controlled by complex internal bed structure and associated interactions with the gas and liquid flows. The first step in estimating the hydrodynamic parameters is estimation of prevailing flow regimes under the operating conditions. The correlations and discussion pre-sented in *Flow Regimes* are useful for selection of criteria to evaluate the flow regime and its transition so that all further calculations will be based on characteristics of that flow regime. Trickle bed reactors are often operated near the boundary of trickle and pulse flow regimes in practice. It is therefore important to be able to estimate the transition boundary between trickle and pulse regimes accurately. The methods and models presented in *Flow Regime Transition* can be used to determine transition boundary. Methods and corre-lations for estimating other key hydrodynamic parameters such as pressure drop, liquid holdup, mass and heat transfer coefficients, and axial dispersion are discussed in *Estimation of Key Hydrodynamic Parameters.*

It is important to note that the hydrodynamics of trickle beds is a complex function of interactions of particle properties, packing characteristics of the bed, properties of gas and liquid, and operating conditions. Therefore, literature data when correlated for particular parameters give considerably different values for same operating parameters. Error bar for experimentally measured values and the predicted ones using various correlations is anywhere between $\pm 15\%$ and $\pm 60\%$. It is often necessary to extrapolate the results of correlations or models to conditions beyond those considered during their development. It is critical in such cases to use multiple models and correlations and compare their extrapolated results to draw appropriate conclusions. Accumulated experience and application of the computational fluid dynamic models (discussed later in Chapter 4) are often used to guide such extrapolation exercise. It is also important to assess the sensitivity of predictions of reactor models and key engineering conclusions with variation of the estimated values of design parameters. Such assessment may provide guidelines for determining desired accuracy and precision required while estimating these design parameters.

The methods, correlations, and models discussed here will be useful to complement the reaction engineering and computational fluid dynamics models of trickle bed reactors discussed in the following chapters.

REFERENCES

Achenbach, E. (1995). Heat and flow characteristics of packed bed. *Experimental Thermal and Fluid Science, 27*, 10–17.

Alicilar, A., Bicer, A., & Murathan, A. (1994). The relation between wetting efficiency and liquid holdup in packed columns. *Chemical Engineering Communications, 128*, 95–107.

Al-Dahhan, M. H., & Dudukovic, M. P. (1994). Pressure drop and liquid holdup in high pressure trickle-bed reactors. *Chemical Engineering Science, 49*, 5681–5698.

Al-Dahhan, M. H., & Dudukovic, M. P. (1995). Catalyst wetting efficiency in trickle-bed reactors at high pressure. *Chemical Engineering Science, 50*(15), 2377–2389.

Attou, A., & Ferschneider, G. (2000). A two-fluid hydrodynamic model for the transition between trickle and pulse flow in a co-current gas–liquid packed-bed reactor. *Chemical Engineering Science, 5*, 511–5491.

Baker, O. (1954). Simultaneous flow of oil and gas. *Oil and Gas Journal, 53*, 185.

Bansal, A., Wanchoo, R. K., & Sharma, S. K. (2005). Flow regime transition in a trickle bed reactor. *Chemical Engineering Communications, 192*, 1046–1066.

Baussaron, L., Julcour-Lebigue, C., Wilhelm, A., Delmas, H., & Boyer, C. (2007). Wetting topology in trickle bed reactors. *AIChE Journal, 53*(7), 1850–1860.

Benkrid, K., Rode, S., & Midoux, N. (1997). Prediction of pressure drop and liquid saturation in trickle-bed reactors operated in high interaction regimes. *Chemical Engineering Science, 52*, 4021–4032.

Boelhouwer, J. G., Piepers, H. W., & Drinkenburg, A. A. H. (2001). Particle–liquid heat transfers in trickle-bed reactors. *Chemical Engineering Science, 56*, 1181–1187.

Boelhouwer, J. G., Piepers, H. W., & Drinkenburg, A. A. H. (2002). Nature and characteristics of pulsing flow in trickle-bed reactors. *Chemical Engineering Science, 57*, 4865–4876.

Briens, C. L., Del Pozo, M., Trudell, C., & Wild, G. (1999). Measurement and modeling of particle–liquid heat transfer in liquid–solid and gas–liquid–solid fluidized beds. *Chemical Engineering Science, 54*, 731–739.

Buffman, B. A., Gibilaro, L. G., & Rathor, M. N. (1970). A probabilistic time delay description of flow in packed beds. *AIChE Journal, 16*, 218–223.

Buffham, B. A., & Rathor, M. N. (1978). The influence of viscosity on axial mixing in trickle flow in packed bed. *Transactions of Institute of Chemical Engineering,* London, *56*, 266.

Burghardt, A., Bartelmus, G., Jaroszynski, M., & Kolodziej, A. (1995). Hydrodynamics and mass transfer in a three-phase fixed-bed reactor with cocurrent gas–liquid downflow. *Chemical Engineering Journal and the Biochemical Engineering Journal, 58*(2), 83.

Burghardt, A., Bartelmus, G., & Szlemp, A. (2004). Hydrodynamics of pulsing flow in three-phase fixed-bed reactor operating at an elevated pressure. *Industrial and Engineering Chemistry Research, 43*, 4511–4521.

Burghardt, A., Kolodziej, A., & Jaroszynski, M. (1990). Experimental studies of liquid–solid wetting efficiency in trickle bed cocurrent reactors. *Chemical Engineering and Processing, 18*, 35–49.

Cassanello, M., Larachi, F., Laurent, A., Wild, G., & Midoux, N. (1996). Gas—liquid mass transfer in high pressure trickle-bed reactors: experiments and modeling. In Ph. Rudolf von Rohr, & Ch. Trepp (Eds.), *Proceedings of the third international symposium on high pressure chemical engineering, Zurich, Switzerland, Oct 7—9, Vol. 12* (pp. 493). Amsterdam, The Netherlands: Elsevier B.V.

Cassanello, M. C., Martinez, O. M., & Cukierman, A. L. (1992). Effect of the liquid axial dispersion on the behavior of fixed bed three phase reactors. *Chemical Engineering Science, 47*(13—14), 3331—3338.

Charpentier, J.-C., Bakos, M., & Le Goff, P. (1971). Hydrodynamics of two-phase concurrent downflow in packed-bed reactors. Gas—liquid flow regimes. Liquid axial dispersion and dead zones. *Proceedings of the Second Congress Applied Physical Chemistry, 2,* 31—47.

Charpentier, J. C., & Favier, M. (1975). Some liquid holdup experimental data in trickle-bed reactors for foaming and nonfoaming hydrocarbons. *AIChE Journal, 21,* 1213—1218.

Charpentier, J. C., Prost, C., & Le Goff, P. (1969). NonEnChute de Pression pour des écoulements à cocourant dans des colonnes à garnissage arrosé: comparaison avec le garnissage noyé. *Chemical Engineering Science, 24,* 1777—1794.

Chhabra, R. P., Comiti, J., & Machac, I. (2001). Flow of non-Newtonian fluids in fixed and fluidized bed. *Chemical Engineering Science, 56,* 1—27.

Chou, T. S., Worley, F. L., & Luss, D., Jr. (1977). Transition to pulsed flow in mixed-phase cocurrent downflow through a fixed bed. *Industrial and Engineering Chemistry Process Design and Development, 16,* 424—427.

Chou, T. S., Worley, F. L., Jr., & Luss, D. (1979). Local particle—liquid mass transfer fluctuations in mixed-phase co-current down flow through a fixed bed in the pulsing regime. *Industrial and Engineering Chemistry Fundamentals, 18,* 279.

Chu, C. F., & Ng, K. M. (1985). Effective thermal conductivity in trickle-bed reactors: application of effective medium theory and random walk analysis. *Chemical Engineering Communications, 37,* 127—140.

Clements, L. D. (1978). Dynamic liquid holdup in cocurrent gas—liquid downflow in packed beds, 1. In T. N. Veziroglu, & S. Kakac (Eds.), *Phase transport and reactor safety.* Washington: Hemisphere Publ. Corp.

Clements, L. D., & Schmidt, P. C. (1980). Dynamic liquid holdup in two-phase downflow in packed beds: air—silicone oil systems. *AIChE Journal, 26,* 317—319.

Colombo, A. J., Baldi, G., & Sicardi, S. (1976). Solid—liquid contacting effectiveness in trickle-bed reactors. *Chemical Engineering Science, 31,* 1101.

Danckwerts, P. V. (1953). Continuous flow systems. Distribution of residence times. *Chemical Engineering Science, 2,* 1—13.

Delaunay, C. B., Storck, A., Laurent, A., & Charpentier, J. C. (1982). Electrochemical determination of liquid—solid mass transfer in a fixed bed irrigated gas—liquid reactor with downward co-current flow. *International Chemical Engineering, 22*(2), 244—251.

Duduković, M. P. (1977). Catalyst effectiveness factor and contacting efficiency in trickle-bed reactors. *AIChE Journal, 23,* 940—944.

Dudukovic, M. P. (2000). Opaque multiphase reactors: experimentation, modeling and troubleshooting. *Oil & Gas Science and Technology — Rev. IFP, 55*(2), ,135—158.

Dudukovic, M. P., & Mills, P. L. (1986). Contacting and hydrodynamics in trickle-bed reactors. In N. P. Cheremisinoff (Ed.), *Encyclopedia of fluid mechanics: gas—liquid flows, Vol. 3* (pp. 969). Houston, TX: Gulf Publ. Corp.

Dwivedi, P. N., & Upadhyah, S. N. (1977). Particle-fluid mass transfer in fixed and fluidized beds. *Industrial and Engineering Chemistry Process Design and Development, 16*(2), 157.

El-Hisnawi, A. (1981). *Tracer and Reaction Studies in Trickle-Bed Reactors. D.Sc. Thesis.* St. Louis, USA: Washington Univ.

El-Hisnawi, A. A., Dudukovic, M. P., & Mills, P. L. (1982). Trickle-bed reactors: dynamic tracer tests, reaction studies and modelling of reactor performance. *ACS Symposium Series, 196,* 421−440.

Ellman, M. J., Midoux, N., Laurent, A., & Charpentier, J. C. (1988). A new improved pressure drop correlation for trickle bed reactors. *Chemical Engineering Science, 43,* 1677−1684.

Ellman, M. J., Midoux, N., Wild, G., Laurent, A., & Charpentier, J. C. (1990). A new improved liquid hold-up correlation for trickle-bed reactors. *Chemical Engineering Science, 45,* 1677−1684.

Fu, M. S., & Tan, C. S. (1996). Liquid holdup and axial dispersion in trickle-bed reactors. *Chemical Engineering Science, 24,* 5357−5361.

Fukushima, S., & Kusaka, K. (1977a). Interfacial area and boundary of hydrodynamic flow region in packed column with cocurrent downward flow. *Journal of Chemical Engineering of Japan, 10,* 461.

Fukushima, S., & Kusaka, K. (1977b). Liquid-phase volumetric and mass transfer coefficient, and boundary of hydrodynamic flow region in packed column with cocurrent downward flow. *Journal of Chemical Engineering of Japan, 10,* 468.

Germain, A. H., Lefebvre, A. G., L'Homme, G. A. (1974). Experimental study of a catalytic trickle-bed reactor. In, Chemical Reaction Engineering II, Advances in Chemistry Series No. 133, ACS, 164 (Chapter 13).

Gianetto, A., Baldi, G., & Specchia, V. (1970). Absorption in packed towers with concurrent high velocity flows: interfacial areas. *Quaderni Dell Ingegnere Chimico Italino, 6,* 125.

Gianetto, A., Specchia, V., & Baldi, G. (1973). Absorption in packed towers with concurrent downward high velocity flows − II: mass transfer. *AIChE Journal, 19,* 916.

Goto, S., & Smith, J. M. (1975). Trickle-bed reactor performance: I − holdup and mass transfer effects. *AIChE Journal, 21,* 706.

Grosser, K., Carbonell, R. G., & Sundaresan, S. (1988). Onset of pulsing in two-phase cocurrent downflow through a packed bed. *AIChE Journal, 34,* 1850−1860.

Gunjal, P. R., Ranade, V. V., & Chaudhuri, R. V. (2003). Liquid-phase residence time distribution in trickle bed reactors: experiments and CFD simulations. *The Canadian Journal of Chemical Engineering, 81,* 821.

Gunjal, P. R., Kashid, M. N., Ranade, V. V., & Chaudhuri, R. V. (2005). Hydrodynamics of trickle bed reactors: experiments and CFD modeling. *Industrial and Engineering Chemistry Research, 44,* 6278−6294.

Hanika, J. (1999). Safe operation and control of trickle-bed reactor. *Chemical Engineering Science, 54,* 4653−4659.

Hashimoto, K., Muroyama, K., Fujiyoshi, K., & Nagata, S. (1976). Effective radial thermal conductivity in cocurrent flow of a gas and liquid through a packed bed. *International Chemical Engineering, 16,* 720−727.

Herskowitz, M., Carbonell, R. G., & Smith, J. M. (1979). Effectiveness factors and mass transfer in trickle-bed reactors. *AIChE Journal, 25,* 272.

Herskowitz, M., & Smith, J. M. (1983). Trickle bed reactors a review. *AIChE Journal, 29,* 1.

Highfill, W., & Al-Dahhan, M. (2001). Liquid−solid mass transfer coefficient in high pressure trickle bed reactors. *Chemical Engineering Research and Design, 79,* 631.

Hirose, T., Mori, Y., & Sato, Y. (1976). Liquid-to-particle mass transfer in fixed bed reactor with cocurrent gas−liquid flow. *Journal of Chemical Engineering of Japan, 9,* 220.

Hochman, J. M., & Effron, E. (1969). Two-phase cocurrent downflow in packed beds. *Industrial and Engineering Chemistry Fundamentals., 8*, 63.

Holub, R. A., Dudukovic, M. P., & Ramachandran, P. A. (1992). A phenomenological model for pressure drop, liquid holdup, and flow regime transition in gas–liquid trickle flow. *Chemical Engineering Science, 47*, 2343–2348.

Holub, R. A., Dudukovic, M. P., & Ramachandran, P. A. (1993). Pressure drop, liquid holdup, and flow regime transition in trickle flow. *AIChE Journal, 39*, 302–321.

Huang, X., Varma, A., & McCready, M. J. (2004). Heat transfer characterization of gas–liquid flows in a trickle-bed. *Chemical Engineering Science, 59*, 3767–3776.

Iliuta, I., & Larachi, F. (1999). The generalized slit model: pressure gradient, liquid holdup and wetting efficiency in gas–liquid trickle flow. *Chemical Engineering Science, 54*, 5039–5045.

Iliuta, I., & Thyrion, F. C. (1997). Gas–liquid mass transfer in fixed beds with two-phase cocurrent downflow: gas Newtonian and non-Newtonian liquid systems. *Chemical Engineering and Technology, 20*, 538–549.

Iliuta, I., Thyrion, F. C., & Muntean, O. (1996). Hydrodynamic characteristics of two-phase flow through fixed beds: air/Newtonian and non-Newtonian liquids. *Chemical Engineering Science, 51*, 4987–4995.

Jaffe, S. B. (1976). Hot spot simulation in commercial hydrogenation processes. *Industrial and Engineering Chemistry Process Design and Development, 15*, 410–416.

Jolls, K. R., & Hanratty, T. J. (1966). Transition to turbulence for flow through a dumped bed of spheres. *Chemical Engineering Science, 21*, 1185–1190.

Julcour-Lebigue, C., Baussaron, L., Delmas, H., & Wilhelm, A. (2007). Theoretical analysis of tracer method for the measurement of wetting efficiency. *Chemical Engineering Science, 62* (18–29), 5374.

Kan, K. M., & Greenfield, P. F. (1979). Pressure drop and holdup in two-phase cocurrent trickle flows through beds of small particles. *Industrial and Engineering Chemistry Process Design and Development, 18*, 760.

Kawase, Y., & Ulbrecht, J. (1981). Newtonian fluid sphere with rigid or mobile interface in a shear thinning liquid drag and mass transfer. *Chemical Engineering Communications, 8*, 213–228.

Kobayashi, S., Kushiyama, S., Ida, Y., & Wakao, W. (1979). Flow characteristics and axial dispersion in two-phase downflow in packed columns. *Kagaku Kogaku Ronbunshu, 5*, 256.

Krieg, D. A., Helwick, J. A., Dillon, P. O., & McCready, M. J. (1995). Origin of disturbances in cocurrent gas–liquid packed bed flows. *AIChE Journal, 41*, 1653–1666.

Lakota, A., & Levec, J. (1990). Solid–liquid mass transfer in packed beds with cocurrent downward two-phase flow. *AIChE Journal, 36*, 1444.

Lamine, A. S., Gerth, L., Legall, H., & Wild, G. (1996). Heat transfer in a packed bed reactor with cocurrent down flow of a gas and a liquid. *Chemical Engineering Science, 51*(5), 3813–3827.

Lange, R. (1978). *Beitrag zur experimentellen Untersuchung und Modellierung von Teilprozessen für katalytische Dreiphasenreaktionen im Rieselbettreaktor. Ph.D. Thesis.* Germany: Technical University of Leuna-Merseburg.

Larachi, F., Cassanello, M., & Laurent, A. (1998). Gas–liquid interfacial mass transfer in trickle-bed reactors at elevated pressures. *Industrial and Engineering Chemistry Research, 37*, 718–733.

Larachi, F., Laurent, A., Midoux, N., & Wild, G. (1991). Experimental study of a trickle-bed reactor operating at high pressure: two-phase pressure drop and liquid saturation. *Chemical Engineering Science, 46*, 1233–1246.

Larachi, F., Laurent, A., Wild, G., & Midoux, N. (1993). Effect of pressure on the trickle-pulsed transition in irrigated fixed bed catalytic reactors. *Canadian Journal of Chemical Engineering, 71*, 319.

Latifi., M. A., Laurent, A., & Storck, A. (1988). Liquid−solid mass transfer in a packed bed with downward cocurrent gas−liquid flow: an organic liquid phase with high Schmidt number. *Chemical Engineering Journal, 34*, 47.

Latifi, M. A., Midoux, N., Storck, A., & Gence, J. M. (1989). The use of microelectrodes in the study of flow regimes in a packed bed reactor with single phase liquid flow. *Chemical Engineering Science, 44*, 2501−2508.

Latifi, M. A., Naderifar, A., Midoux, N., & Le Mehaute, A. (1994). Fractal behavior of local liquid−solid mass transfer fluctuation at the wall of a trickle bed reactor. *Chemical Engineering Science, 49*, 3823−3829.

Llano, J. J., Rosal, R., Sastre, H., & Díez, F. V. (1997). Determination of wetting efficiency in trickle-bed reactors by a reaction method. *Industrial and Engineering Chemistry Research, 36*, 2616−2625.

Lutran, P. G., Ng, K. M., & Delikat, E. P. (1991). Liquid distribution in trickle beds. An experimental study using computer-assisted tomography. *Industrial and Engineering Chemistry Research, 30*, 1270−1280.

Marcandelli, C., Wild, G., Lamine, A. S., & Bernard, J. R. (1999). Measurement of local particle−fluid heat transfer coefficient in trickle-bed reactors. *Chemical Engineering Science, 54*, 4997−5002.

Matsuura, A., Hitaka, Y., Akehata, T., & Shirai, T. (1979). Effective radial thermal conductivity in packed beds with downward cocurrent gas−liquid flow. *Heat Transfer Japanese Research, 8*, 44−52.

Matsuura, Akinori, Takashi, Akehata, & Takashi, Shirai (1976). Axial dispersion of liquid in concurrent gas−liquid downflow in packed beds. *Journal of Chemical Engineering of Japan, 9*(4), 294−301.

Mears, D. E. (1971). The role of axial dispersion in trickle flow laboratory reactions. *Chemical Engineering Science, 26*, 1361.

Michell, R. W., & Furzer, I. A. (1972). Mixing in trickle flow through packed beds. *Chemical Engineering Journal, 4*, 53.

Midoux, N., Favier, M., & Charpentier, J. C. (1976). Flow pattern, pressure loss and liquid holdup data in gas−liquid downflow packed beds with foaming and nonfoaming hydrocarbons. *Journal of Chemical Engineering of Japan, 9*, 350−356.

Midoux, N., Morsi, B. I., Purwasmita, M., Laurent, A., & Charpentier, J. C. (1984). Interfacial area and liquid-side mass transfer coefficient in trickle-bed reactors operating with organic liquids. *Chemical Engineering Science, 39*, 781.

Mills, P. L., & Duduković, M. P. (1981). Evaluation of liquid−solid contacting in trickle-bed reactors by tracer methods. *AIChE Journal, 27*, 893−904.

Morsi, B. I., Midoux, N., & Charpentier, J. C. (1978). Flow patterns and some holdup experimental data in trickle-bed reactors for foaming, nonfoaming and viscous organic liquids. *AIChE Journal, 24*, 357−360.

Morsi, B. I., Midoux, N., Laurent, A., & Charpentier, J. C. (1982). Hydrodynamics and interfacial areas in downward cocurrent gas−liquid flow through fixed beds: influence of the nature of the liquid. *International Chemical Engineering, 22*, 142−151.

Muroyama, K., Hashimoto, K., & Tomita, T. (1977). Heat transfer from the wall in gas−liquid cocurrent packed beds. *Kagaku Kogaku Ronbunshu, 3*, 612−616.

Nelson, P. A., & Galloway, T. R. (1975). Particle-to-fluid heat and mass transfer in dense systems of fine particles. *Chemical Engineering Science, 30*, 11–16.

Ng, K. M. (1986). A model for regime transitions in cocurrent down-flow trickle-bed reactors. *AIChE Journal, 32*, 115–122.

Otake, T., & Kunugita, E. (1958). Mixing characteristics of irrigated packed towers. *Chem. Eng. Japan., 22*, 144–150.

Pant, H. J., Saroha, A.K., & Nigam, K. D. P. (2000). Measurement of liquid holdup and axial dispersion in trickle bed reactors using radiotracer technique. *NUKLEONIK, 45*(4), 235–241.

Ramachandran, P. A., & Smith, J. M. (1979). Mixing-cell method for design of trickle-bed reactors. *Chemical Engineering Journal, 17*(2), 91–99.

Ranz, W. E. (1952). Friction and heat transfer co-efficient for single particles and packed beds. *Chemical Engineering Progress, 48*(8), 247.

Rao, V. G., Ananth, M. S., & Varma, Y. B. G. (1983). Hydrodynamics of two-phase cocurrent downflow through packed beds (parts I and II). *AIChE Journal, 29*, 467–483.

Rao, V. G., & Drinkenburg, A. A. H. (1985). Solid–liquid mass transfer in packed beds with cocurrent gas–liquid down flow. *AIChE Journal, 31*, 1059.

Ring., Z. E., & Missen, R. W. (1991). Trickle-bed reactors: tracer study of liquid holdup and wetting efficiency at high temperature and pressure. *Canadian Journal of Chemical Engineering, 69*, 1016.

Saez, A. E., & Carbonell, R. G. (1985). Hydrodynamic parameters for gas–liquid cocurrent flow in packed beds. *AIChE Journal, 31*, 52–62.

Sai, P. S. T., & Varma, Y. B. G. (1987). Pressure drop in gas–liquid downflow through packed beds. *AIChE Journal, 33*, 2027–2036.

Sai, P. S. T., & Varma, Y. B. G. (1988). Flow pattern of the phases and liquid saturation in gas–liquid cocurrent downflow through packed beds. *Canadian Journal of Chemical Engineering, 66*, 353–360.

Saroha, A. K., & Nigam, K. D. P. (1996). Trickle bed reactor. *Reviews in Chemical Engineering, 12*(3), 207–346.

Saroha, A. K., Nigam, K. D. P., Saxena, A. K., & Kapoor, V. K. (1998). Liquid distribution in trickle-bed reactors. *AIChE Journal, 44*, 2044–2052.

Satterfield, C. N. (1975). Trickle bed reactors. *AIChE Journal, 21*, 209–228.

Satterfield, C. N., Van Eek, M. W., & Bliss, G. S. (1978). Liquid-solid mass transfer in Packed Beds with downward concurrent gas-liquid flow. *AIChE J., 24*, 709.

Scheidegger, A. E. (1974). *The Physics of Flow through Porous Media* (3rd ed.). univ. de Toronto Press.

Schwartz, J. G., Weger, E., & Dudukovic, M. P. (1976). A new tracer method for determination of solid–liquid contacting efficiency in trickle-bed reactors. *AIChE Journal, 22*, 894.

Sederman, A. J., & Gladden, L. F. (2001). Magnetic resonance imaging as a quantitative probe of gas–liquid distribution and wetting efficiency in trickle-bed reactors. *Chemical Engineering Science, 56*, 2615–2628.

Seguin, D., Montillet, A., & Comiti, J. (1998). Experimental characterization of flow regimes in various porous media – I: limit of laminar flow regime. *Chemical Engineering Science, 53*(21), 3751.

Seguin, D., Montillet, A., Comiti, J., & Huet, F. (1998). Experimental characterization of flow regimes in various porous media – II: transition to turbulent regime. *Chemical Engineering Science, 53*, 3897–3909.

Sicardi, S., Gerhard, H., & Hofmann, H. (1979). Flow regime transition in trickle-bed reactors. *Chemical Engineering Journal, 18*, 173−182.

Sicardi, S., & Hofmann, H. (1980). Influence of gas velocity and packing geometry on pulsing inception in trickle bed reactors. *Chemical Engineering Journal, 20*, 251−253.

Sie, S. T., & Krishna, R. (1998). Process development and scale up − III. Scale-up and scale-down of trickle bed processes. *Reviews in Chemical Engineering, 14*, 203.

Specchia, V., & Baldi, G. (1977). Pressure drop and liquid holdup for two phase concurrent flow in packed beds. *Chemical Engineering Science, 32*, 515−523.

Specchia, V., & Baldi, G. (1979). Heat transfer in trickle-bed reactors. *Chemical Engineering Communications, 3*, 483−499.

Specchia, V., Baldi, G., & Gianetto, A. (1978). Solid−liquid mass transfer in concurrent two-phase flow through packed beds. *Industrial and Engineering Chemistry Process Design and Development, 17*, 362.

Talmor, E. (1977). Two-phase downflow through catalyst beds. Parts I and II. *AIChE Journal, 23*, 868−878.

Tan, C. S., & Smith, J. M. (1982). A dynamic method for liquid-solid mass transfer in Trickle Beds. *AIChE J., 28*, 190.

Tosun, G. (1982). Axial *dispersion* in trickle bed reactors: influence of the gas flow rate. *Industrial and Engineering Chemistry Fundamentals, 21*, 184−186.

Tsochatzidis, N. A., & Karabelas, A. J. (1994). Study of pulsing flow in a trickle bed using the electro-diffusion technique. *Journal of Applied Electrochemistry, 24*, 670−675.

Turek, F., Lange, R., & Bush, A. (1979). *Chemie Technik., 31*, 232.

Van Swaaij, W. P. M., Charpentier, J. C., & Villermaux, J. (1969). Residence time distribution in the liquid phase of trickle flow in packed columns. *Chemical Engineering Science, 24*, 1083.

Wammes, W. J. A., & Westerterp, K. R. (1990). The influence of reactor pressure on the hydrodynamics in a cocurrent gas−liquid trickle-bed reactor. *Chemical Engineering Science, 45*, 2247−2254.

Wammes, W. J. A., & Westerterp, K. R. (1991). Hydrodynamics in a pressurized cocurrent gas−liquid trickle-bed reactor. *Chemical Engineering Technology, 14*, 406.

Wanchoo, R. K., Kaur, N., & Bansal, M. (2007). RTD in trickle-bed reactors: experimental study. *Chemical Engineering Communications, 194*(1503), 50−68.

Wang, Y., Mao, Z., & Chen, J. (1998). A new instrumentation for measuring the small scale maldistribution of liquid flow in trickle beds. *Chemical Engineering Communications, 163*, 233−244.

Weekman, V. W. (1976). Hydroprocessing reaction engineering. In *Proceedings of the fourth international/sixth European symposium on chemical reaction engineering*. Heidelberg, Germany: VDI.

Weekman, V. W., & Myers, J. E. (1964). Fluid-flow characteristics of concurrent gas−liquid flow in packed beds. *AIChE Journal, 10*, 951−957.

Weekman, V. W., & Myers, J. E. (1965). Heat transfer characteristics of concurrent gas−liquid flow in packed beds. *AIChE Journal, 11*, 13−17.

Wild, G., Larachi, F., & Charpentier, J.-C. (1992). Heat and mass transfer in gas−liquid−solid fixed bed reactors. In M. Quintard, & M. Todorović (Eds.), *Heat and mass transfer in porous media*. Amsterdam, The Netherlands: Elsevier. p. 616.

Wu, Y., Khadilkar, M. R., Al-Dahhan, M. H., & Dudukovic, M. P. (1996). Comparison of upflow and downflow two-phase flow packed bed reactors with and without fines: experimental observations. *Industrial and Engineering Chemical Research, 35*(2), 397−405.

Chapter 3

Reaction Engineering of Trickle Bed Reactors

Therefore O students, study mathematics and do not build without foundations.

Leonardo Da Vinci

INTRODUCTION

Trickle bed reactors represent one of the most commonly used industrial reactors for performing chemical processes which involve solid-catalyzed reactions of gas and liquid phase reactants. Key technological aspects of trickle bed reactor and some examples of industrial trickle bed processes are discussed in Chapter 1. Trickle bed reactor can be described as a fixed bed of catalyst particles with cocurrent or countercurrent flow of gas and liquid phase reactants to produce gas and/or liquid phase products. However, in most industrial operations, downflow mode is practiced. They are preferred due to simple design, low pressure drop (compared to liquid full operation), reasonable heat transfer efficiency, and convenience of using interstage quenching to control temperature. The various modes of operation of the multiphase fixed-bed reactors are shown in Fig. 1. The trickle flow operation is most commonly used in large volume industrial productions like hydroprocessing, hydration, amination, and oxidation reactions.

The trickle flow operation is advantageous in many chemical processes to achieve high conversion efficiency in the production of chemicals or removal of pollutants from gas or liquid feeds. When compared with vapor phase catalytic reactors, the presence of a liquid phase is a major difference which adds complexities of interphase mass and heat transfer as well as hydrodynamic flow regimes that can exist depending on the operating gas and liquid velocities.

Flow regimes and other hydrodynamic characteristics of trickle bed reactors (discussed in Chapter 2) primarily affect the external mass and heat transfer coefficients, axial and radial dispersion coefficients, the uniformity of liquid flow through the bed and external wetting efficiency of the catalyst particles and bed. In this chapter, reaction engineering aspects of trickle bed

Trickle Bed Reactors. DOI: 10.1016/B978-0-444-52738-7.10003-8

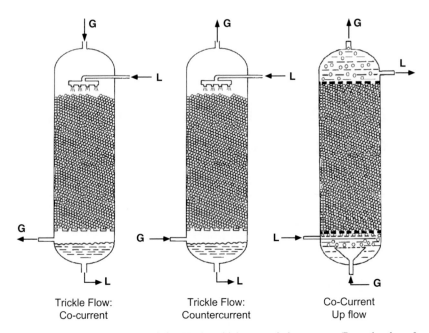

Trickle Flow: Trickle Flow: Co-Current
Co-current   Countercurrent   Up flow

FIGURE 1   Operating modes of fixed-bed multiphase catalytic reactors (Ramachandran & Chaudhari, 1983).

reactor are discussed to address the basic principles of the rate and reactor performance analysis. These would be similar to fixed-bed catalytic reactors but with differences with respect to presence of liquid phase (and it's interaction with gas and solid phases) in the trickle bed reactor. It is necessary to understand the rate analysis on a particle scale as well as the reactor scale; the latter being more dependent on our understanding of the hydrodynamics and mass transfer properties of the fluid phases (gas and liquid). The analysis of rate and reactor performance for solid-catalyzed multiphase reactions of gas and liquid phase reactants has been well developed over the years as documented in reviews by Chaudhari and Ramachandran (1980), Dudukovic, Larachi, and Mills (1999), Gianetto and Silveston (1986), Satterfield (1975), and monographs by Ramachandran and Chaudhari (1983) and Shah (1979). Some of the early developments were based on oversimplified assumptions of the reaction kinetics (e.g., first order with gas and zero order with liquid reactant: Satterfield, 1975; Shah, 1979), which is not true for most of the practical examples in industry. A significant advance in this direction was due to Ramachandran and Chaudhari (1983), who developed approaches for analysis of multiphase reactions with non-linear kinetics including Langmuir–Hinshelwood-type mechanisms and also for complex multistep reactions. Trickle bed reactors have some unique features and complexities which demand careful account of the various aspects of hydrodynamics, mass transfer, and reaction kinetics to be able to develop predictive models for their performance and understand the guidelines for design and scale-up.

In order to understand the rate processes and performance of a trickle bed reactor, it is important to first reflect on the operating features and important factors that influence the overall performance of a process. Some of these factors are scale dependent and some are not. It is useful to consider these separately and then see how these can be used together to evolve the models of trickle bed reactors. Based on current state of understanding, the following factors are important:

- Homogeneity of the bed, liquid distribution, and geometric factors
- Operating modes: cocurrent or countercurrent flow of gas and liquid phases, recycle streams, wetting of catalyst particles, flow distribution (channeling, bed irrigation), axial and radial mixing
- Mass transfer: gas—liquid, fluid—particle (both to wetted and unwetted surfaces of catalyst), and intraparticle diffusion within the porous catalyst on a particle scale or differential conditions and reactor scale
- Reaction kinetics: rate equations (intrinsic kinetics), catalyst deactivation
- Non-isothermal effects: exothermic or endothermic reactions, solvent/ reactant vaporization, multiplicity, and oscillatory behavior

Since, the reactor performance will depend on multiple factors listed above; it is useful to employ various criteria to assess the significance of these effects before a representative reactor model is developed for a specific process. In this chapter, the current state of development on analysis of trickle bed reactors is discussed to address the influence of the above mentioned factors. The discussion is divided into two main sections. The following section discusses the analysis of effective reaction rate. The overall reactor models are discussed in *Reactor Performance Models for Trickle Bed Reactors*. These models with appropriate hydrodynamic inputs (from Chapter 2) will provide a methodology and a platform for simulation of trickle bed reactors.

OVERALL RATE OF REACTION

In this section, general concepts and methodologies of rate analysis are discussed instead of concentrating on specific reactions and processes. From the point of view of rate analysis at a particle level, it is useful to consider two situations prevailing in a trickle bed reactor: (a) complete wetting of catalyst particles and (b) partially wetted catalyst particles. These two situations will largely depend on factors such as gas and liquid velocities and resulting flow regimes, particle geometry (size and shape), liquid distribution, and solvent/reactant vaporization (see *Wetting of the Catalyst Particles* in Chapter 2).

Completely Wetted Catalyst Particles

Consider a situation at higher liquid velocities (>0.5 cm/s), where the catalyst particles are generally wetted completely and the liquid flows over the catalyst particles with a rippling or pulsing flow. In this regime, the interaction between gas and liquid phases is very high and is often referred as "high interaction regime." The rate at the catalyst particle level is represented by equations similar to those for three-phase catalytic reactions except that the mass transfer coefficients will be dependent on the trickle bed hydrodynamics and operating conditions. For a bimolecular reaction of the type $A(g) + vB(l) \rightarrow$ products, the various steps of mass transfer and surface reactions that describe the overall rate of the reaction are (Ramachandran and Chaudhari, 1983) as follows:

(i) Transport of gas phase reactant A from gas phase to the bulk liquid
(ii) Transport of A and B from bulk liquid to the catalyst surface
(iii) Intraparticle diffusion of A and B within the pores of catalyst
(iv) Adsorption of A and B on the catalyst sites and surface chemical reaction of adsorbed A and B to yield products
(v) Desorption of products to the bulk liquid phase

These steps are shown schematically in a concentration profile in Fig. 2.

A detailed analysis of rate equations has been developed by Chaudhari and Ramachandran (1980) and Ramachandran and Chaudhari (1983) for the overall rate of reaction considering different types of reaction kinetics and the influence of both external and intraparticle diffusion effects. In this analysis the concepts of "catalytic effectiveness factor" and "overall effectiveness factor" have been used. The "concept of overall effectiveness factor" allows quantitative account of the influence of both the external and the intraparticle mass transfers and is very convenient in the reactor performance analysis.

For a general case of a reaction of the type $A(g) + v B(l) \rightarrow$ products, the overall effectiveness factor is defined as "the ratio of the actual rate of reaction to the rate in the absence of all the mass transfer effects."

$$\eta_0 = \frac{R_A}{wk(A^*)^m B_1^n} \tag{1}$$

If the order of reaction with respect to reactant A is "m" and with respect to reactant B is "n," the overall effectiveness factor is given by the following expression (Chaudhari & Ramachandran, 1980):

$$\eta_0 = \frac{[1 - (\eta_0/\sigma_A)]}{\phi_0} \left[\coth\left\{ 3\phi_0 \left(1 - \frac{\eta_0}{\sigma_A} \right)^{(m-1/2)} \right\} - \frac{1}{3\phi_0(1 - (\eta_0/\sigma_A))^{(m-1)/2}} \right] \tag{2}$$

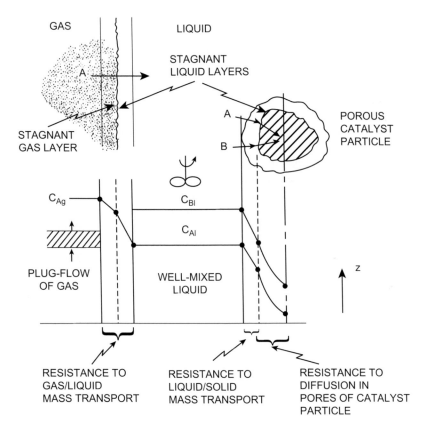

FIGURE 2 Concentration profiles for a three-phase catalytic reaction under differential conditions (Mills & Chaudhari, 1997).

where, σ_A and ϕ_0, the parameters representing external and intraparticle mass transfer contributions, respectively, are given by:

$$\sigma_A = \frac{A^* \left[\frac{1}{k_L a} - \frac{1}{k_s a_p} \right]^{-1}}{wk(A^*)^m B_l^n} \tag{3}$$

and

$$\phi_0 = \frac{R}{3} \left[\frac{(m+1)S_p k(A^*)^{m-1} B_l^n}{2D_e} \right] \tag{4}$$

The overall rate of reaction is then given as:

$$R_A = \eta_0 wk(A^*)^m B_l^n \tag{5}$$

The intrinsic kinetics of the reactions is often represented by more complex rate equations, based on Langmuir—Hinshelwood (L—H) type of models in which additional parameters like adsorption equilibrium constants are involved. For these non-linear rate forms, Ramachandran and Chaudhari (1983) have developed analytical rate forms of effectiveness factors using the concept of generalized Thiele modulus originally proposed by Bischoff (1965). Typical plots of overall effectiveness factor as a function of Thiele modulus for a half-order and L—H-type reaction kinetics are shown in Figs. 3 and 4, respectively.

For complex reactions with consecutive or parallel steps the problem of mass transfer analysis is often more complex and analytical expressions are not possible. Merchan, Emig, Hofmann, and Chaudhari (1986) have developed a mathematical model for three-phase catalytic reactions with consecutive steps to evaluate the overall effectiveness factor incorporating the contributions of external and intraparticle mass transfers as well as heat effects. Both, power law-type and L—H type rate equations have been considered. The reaction scheme considered was

$$\nu_{a1}A(g) + \nu_{b1}B(l) \rightarrow \nu_{c1}C(l)$$

$$\nu_{a2}A(g) + \nu_{c2}C(l) \rightarrow \nu_{e1}E(l) \tag{6}$$

The corresponding rate equations were

$$r_1 = k_{10} \exp\left[-\gamma_1\left(\frac{1}{\theta} - 1\right)\right] A^{*m} B_1^n a^m b^n \tag{7}$$

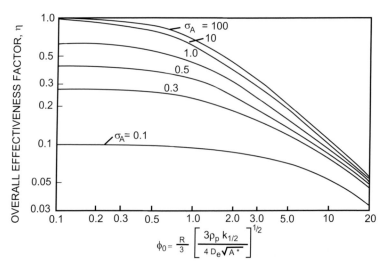

FIGURE 3 Overall effectiveness factor plots for half-order reaction (Ramachandran & Chaudhari, 1983).

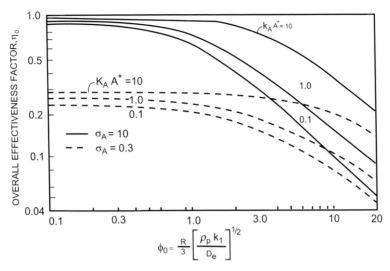

FIGURE 4 Overall effectiveness factor plots for L–H kinetics (Ramachandran & Chaudhari, 1983).

$$r_2 = k_{20} \exp\left[-\gamma_2 \left(\frac{1}{\theta} - 1 \right) \right] A^{*p} B_1^q a^p b^q \tag{8}$$

For the case of L–H kinetics, following rate forms were considered:

$$r_1 = \frac{k_{10} A^* B_1 \exp\left[-\gamma_1 \left(\frac{1}{\theta} - 1 \right) \right] ab}{Den^u} \tag{9}$$

$$r_2 = \frac{k_{20} A^* B_1 \exp\left[-\gamma_2 \left(\frac{1}{\theta} - 1 \right) \right] ac}{Den^u} \tag{10}$$

with

$$Den = 1 + k_a a \exp\left[(-\gamma_3 + \gamma_5) \left(\frac{1}{\theta} - 1 \right) \right] \\ + (k_b b + k_c c) \exp\left[-\gamma_3 \left(\frac{1}{\theta} - 1 \right) \right] \tag{11}$$

The mass and energy balance equations are

$$\frac{d^2 a}{dy^2} + \frac{2}{y} \frac{da}{dy} - \psi_1 - \psi_2 = 0 \tag{12}$$

$$\frac{d^2 c}{dy^2} + \frac{2}{y} \frac{dc}{dy} - q_{c1} \psi_1 - q_{c2} \psi_2 = 0 \tag{13}$$

$$\frac{d^2\theta}{dy^2} + \frac{2}{y}\frac{d\theta}{dy} - \beta_1\psi_1 - \beta_2\psi_2 = 0 \tag{14}$$

With a stoichiometric relation for component B as:

$$b = 1 - \frac{\alpha_b}{\alpha_{c1} + \alpha_{c2}}\left[c - c_0 + \alpha_{c2}\frac{A^*}{B_1}(1 - \alpha)\right] \tag{15}$$

The boundary conditions are

$$y = 0 \quad \frac{da}{dy} = \frac{dc}{dy} = \frac{d\theta}{dy} \tag{16}$$

and

$$y = 1$$

$$\left.\frac{da}{dy}\right|_{y=1} = Sh'_A(1 - a_s) \tag{17}$$

$$-\left.\frac{dc}{dy}\right|_{y=1} = Sh'_C(c_1 - c_s), \quad -\left.\frac{d\theta}{dy}\right|_{y=1} = Bi(1 - \theta_s) \tag{18}$$

The overall effectiveness factor is then defined as:

$$\eta_0 = \frac{a_p D_{\text{eff,A}}\left(\dfrac{dA}{dr}\right)_{r=R}}{\dfrac{w}{\rho_p}(r_1 + r_2)_{\text{bulk conditions}}} \tag{19}$$

Merchan et al. (1986) evaluated overall effectiveness factors for different values of *Sherwood* number ($Sh'_C = k_{sc}R/D_{\text{eff,c}}$ external mass transfer parameter for species C), *Biot* number ($Bi = h_sR/\lambda_{\text{eff}}$ external heat transfer parameter), *Thiele* parameters ($P_1 = R^2(A^*)^{m-1}B_1^n K_{10}/D_{\text{eff,A}}$ and $P_2 = R^2 (A^*)^{p-1}B_1^q K_{20}/D_{\text{eff,A}}$), and thermicity parameter $\beta_1 = (-\Delta H_1)A^* D_{\text{eff,A}}/(T_1\lambda_{\text{eff}})$. Depending on the values of hermicity parameter, multiple solutions were observed for effectiveness factor due to non-isothermal effects. This generalized approach can be used for a variety of practical examples to account for the mass and heat transfer effects in trickle bed reactors. Typical plots of effectiveness factor as a function of Thiele modulus for power law and L–H kinetics are shown in Fig. 5.

The methodology is fairly general and can be applied to any reaction scheme (single or multistep for isothermal or non-isothermal conditions) in a straightforward manner provided the catalyst particles are completely wetted. This analysis provides a rate of reaction at a particle level incorporating

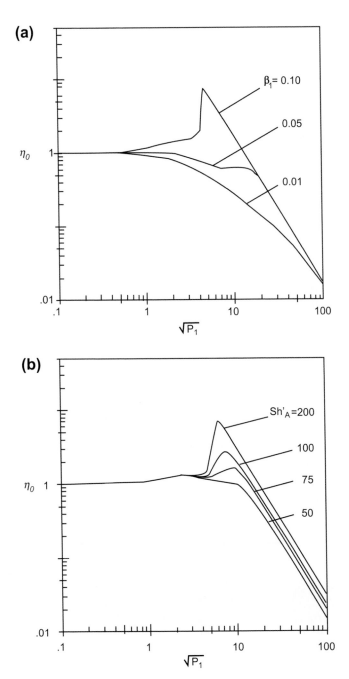

FIGURE 5 Non-isothermal overall effectiveness factor plots for complex consecutive reactions; (a) power law kinetics and [$Sh'_A = 100$, $Bi = 10$, beta2 = beta1] (b) L−H kinetics [$Bi = 10$, beta1 = beta2 = 0.1] (Merchan et al., 1986).

interphase and intraparticle mass and heat transfers coupled with reaction kinetics.

Partially Wetted Catalyst Particles

Many industrial-scale trickle bed processes operate at lower liquid superficial velocities (<0.5 cm/s) at which the catalyst particles are not completely wetted. This phenomenon of incomplete wetting has therefore received significant attention and numerous studies have addressed the influence of wetting on the overall rate, conversion/selectivity behavior, and heat effects in trickle bed reactors. In general, when porous catalyst is used, two types of wetting can exist:

(i) *Internal wetting*: This is the amount of internal area of catalyst particles wetted by the liquid and is a measure of the active surface area available for the reaction. In most cases the internal catalyst is almost completely wetted even if there is incomplete wetting of the external surface due to capillary forces. Under conditions of poor liquid distribution near the entrance of the trickle bed or under conditions of channeling of the liquid flow, some of the catalyst bed is poorly irrigated by the liquid resulting in ineffective internal wetting of the catalyst.

(ii) *External effective wetting*: The phenomena of external wetting are observed to be more significant in many trickle bed processes. This represents the fraction of external area of catalyst wetted by flowing liquid. While, the catalytic reaction occurs over the internal surface area of the particle, inefficient external wetting can significantly influence the overall performance of a trickle bed reactor, especially when the external mass transfer resistance is significant. A schematic illustration of increase in effectiveness factor with wetting efficiency is shown in Fig. 6 (taken from Boelhouwer, Piepers, & Drinkenburg, 2001).

The incomplete wetting conditions also correspond to significant stagnant liquid pockets due to lower liquid superficial velocities as shown schematically

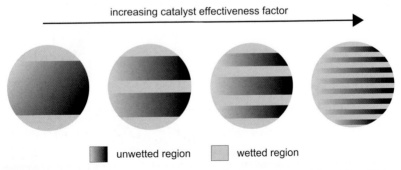

FIGURE 6 Schematic illustration of increase in effectiveness factor with wetting efficiency (Boelhouwer et al., 2001).

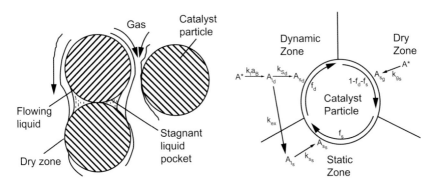

FIGURE 7 Schematic of incomplete wetting and stagnant liquid pockets (from Rajashekharam et al., 1998; Westerterp & Wammes, 2000).

in Fig. 7. Thus, it is important to evaluate the influence of wetting and stagnant pockets on the rate and reactor performance.

In this section, the effect of incomplete wetting on the rate processes in a trickle bed and its performance is discussed. Several models for effectiveness factor analysis have been proposed to describe the influence of partial external wetting of catalyst particles (Goto, Lakota, & Levec, 1981; Herskowitz, Carbonell, & Smith, 1979; Mills & Dudukovic, 1980; Ramachandran & Smith, 1979a; Tan & Smith, 1980). These models predict that the overall rate of reaction increases under conditions of partial wetting for a gas-limited reaction (with significant external mass transfer limitations) due to elimination of the liquid–particle mass transfer resistance for the unwetted part of the catalyst. This phenomenon has also been demonstrated experimentally (Beaudry, Mills, & Dudukovic, 1986; Germain, Lefebvre, & L'Homme, 1974; Herskowitz & Mosseri, 1983; Mills & Dudukovic, 1979; Morita & Smith, 1978; Sedricks & Kenney, 1973). For highly exothermic reactions, evaporation of liquid can lead to local drying of the catalyst surface resulting in incomplete external and internal wetting of catalyst particles. Harold and Ng (1993) investigated this aspect and proposed a theoretical model considering external unwetting of a single catalyst particle for cyclohexene hydrogenation using Pd/C catalyst.

Ramachandran and Smith (1979a) proposed a theoretical model for effectiveness factor considering a rectangular slab of the catalyst particle to represent partial wetting. They not only proposed a numerical solution of the model but also proposed approximate analytical solutions for a first-order reaction assuming that the overall effectiveness factor can be obtained as a weighted sum of the effectiveness factors for totally wetted and totally dry pellets. The generalized expression can be written as:

$$\eta_0 = f_w \eta_w + (1 - f_w)\eta_d \qquad (20)$$

where, η_w is the effectiveness factor of an actively wetted catalyst and η_d is the effectiveness factor for a dry or inactively wetted catalyst pellet. The approximate relations for the overall effectiveness factor for flat plate geometry for the two cases are as follows:

(a) Limiting reactant in the gas phase

$$\eta_0 = \frac{f_w}{\dfrac{\phi}{\tan h\,\phi} + \dfrac{\phi^2}{Sh_{gls}}} + \frac{(1 - f_w)}{\dfrac{\phi}{\tan h\,\phi} + \dfrac{\phi^2}{Sh_{gs}}} \tag{21}$$

(b) Limiting reactant in the liquid phase

$$\eta_0 = \frac{(f_w/\phi)\tan h(\phi/f_w)}{1 + \left(\dfrac{\phi}{Sh_{ls}}\right)\tan h(\phi/f_w)} \tag{22}$$

where,

$$\phi = L\left(\frac{\rho_p k_1}{D_e}\right)^{1/2}, \quad Sh_{gs} = \frac{H_A k_{gs}L}{D_e}, \quad Sh_{gls} = \frac{k_{gls}L}{D_e}, \quad \text{and } Sh_{ls} = \frac{k_s L}{D_e}$$

Lee and Smith (1982) proposed the following criteria for evaluation of the significance of the partial wetting effects for spherical geometry of particles. For the partial wetting effect on the rate to be significant following condition needs to be satisfied:

$$\frac{(d_p^2 \rho_p \mathit{\Omega}_A / 4 D_{eA} A^*)}{\left[1 - \dfrac{\rho_p(1 - \varepsilon_B)\mathit{\Omega}_A}{k_L a_B A^*}\right]} > \frac{(\phi/3)\tan h(\phi/3)}{1 + \dfrac{(\phi/3)\tan h(\phi/3)}{Bi_l}} \tag{23}$$

where,

$$Bi_l = \frac{\varepsilon_p k_L L_r}{3 D_{eA}}$$

These models can be extended to account for the effect of stagnant liquid pockets as described by Chaudhari, Jaganathan, Mathew, Julcour, and Delmas (2002) and Rajashekharam, Jaganathan, and Chaudhari (1998) in which the overall effectiveness factor is extended to incorporate the effect of stagnant liquid holdup as:

$$\eta_0 = f_w \eta_{cd} + f_s \eta_{cs} + (1 - f_w - f_s)\eta_{cg} \tag{24}$$

where, f_s represents the fraction of catalyst surface covered by the stagnant liquid pockets. Subscripts of effectiveness factor cd, cs, and cg indicate dynamic, stagnant, and dry zone, respectively. Depending on the particle size, operating gas and liquid velocities, and scale of operation, the stagnant liquid holdup can be in the range of 10−30% (Koros, 1986).

The methodology discussed here can be used to estimate the effectiveness factor for a generalized reaction scheme under partial wetting conditions. It may be noted that energy liberated during the exothermic reactions on catalyst particles may cause evaporation of solvent and may lead to partial wetting. Exothermic reactions may also influence effectiveness factor significantly due to non-isothermal effects. This is discussed in the following.

Exothermic Reactions

The analysis of non-isothermal effects on the rate of overall reaction in a trickle bed is important since most industrial trickle beds are operated with significant temperature gradients and temperature control is considered as one of the challenges in design and scale-up of these reactors. In addition to the general features of exothermic catalytic reactors associated with solid-catalyzed reactions, additional features due to incomplete wetting, vaporization of liquid phase components, changes in gas solubility, etc., also need to be considered which make the problem more difficult to analyze. Some implications of the exothermic reactions in a trickle bed are summarized here:

- Exothermic reactions lead to significant temperature gradients within a catalyst particle which when coupled with external mass and heat transfer limitations and intraparticle concentration gradients lead to hotspots formation, multiplicity behavior of rates, temperature runaway, hysteresis phenomena, and in some cases deactivation of catalyst and changes in product selectivity.
- Due to large heat effects as a result of exothermicity, one of the reactants or solvent vaporization becomes significant in a trickle bed reactor resulting in incomplete wetting of internal as well as external area of catalyst particles. This, in turn can lead to distortion of liquid distribution patterns, hotspots formation, and increase or decrease of reaction rates depending on reaction kinetics.
- The large temperature gradients within a catalyst particle can also influence the temperature profile and hence, solubility, adsorption equilibrium parameters, and partial pressure of reacting gases reducing the rate of reaction.
- Thus, the exothermic reactions influence multiple parameters which further affect the rate, conversion, and selectivity behavior in a trickle bed reactor and a careful account of this factor is necessary to design optimum reactor operation and safety.

Hanika, Sporka, Ruzicka, and Rstka (1976) showed that for exothermic hydrogenation of cyclohexene, hysteresis of rates is observed besides significant temperature gradients in a trickle bed reactor. Typical results are shown in Figs. 8 and 9 for hydrogenation of cyclohexene illustrating the effect of exothermic reaction and temperature rise.

Increase in hydrogenation rate and the resulting temperature rise leads to vaporization of liquid phase reactant, cyclohexene, resulting in partial or complete drying (unwetting) of the catalyst surface. In this situation, a very rapid gas phase catalytic hydrogenation of vaporized cyclohexene leads to sharp temperature rise and a hysteresis behavior as shown in Fig. 10. Hanika et al. (1976) proposed the following criteria to assess the significance of liquid vaporization in a trickle bed reactor:

$$\frac{\rho_L \mu_l M_G}{\rho_G \mu_G M_L} < \frac{P_v(T)}{P - P_v(T)} \tag{25}$$

M_G and M_L are molecular weights of gas and liquid phases, respectively. The effect of volatility on trickle bed reactor performance has been addressed by several experimental studies under conditions of hydroprocessing of petroleum feed stocks (Akgerman, Collins, & Hook, 1985; Kocis & Ho, 1986; Morris Smith & Satterfield, 1986).

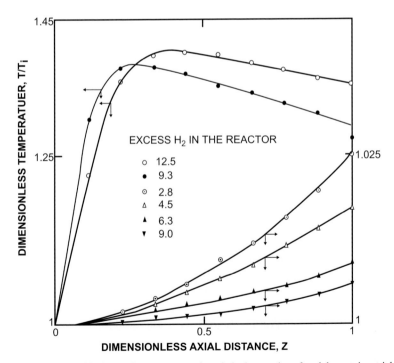

FIGURE 8 Effect of hydrogen flow rate on exothermic hydrogenation of cyclohexene in a trickle bed reactor (Hanika et al., 1976).

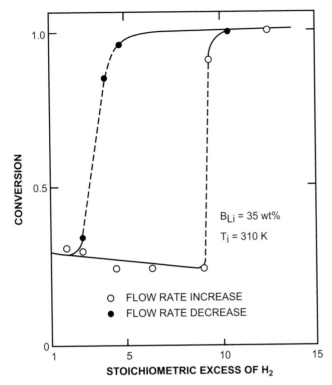

FIGURE 9 Experimental observation of hysteresis behavior of hydrogenation rate in a trickle bed reactor as a result of hydrogen flow rate (Hanika et al., 1976).

Lysova et al. (2007) have used MRI technique to map the liquid distribution effects in catalyst particles and a trickle bed reactor under conditions of exothermic reactions and significant volatility. The existence of the partially wetted pellets in a catalyst bed which are potentially responsible for the appearance of hotspots in the reactor has been visualized (see Fig. 10a). They also observed oscillations of the reactor bed temperature (Fig. 10b) for hydrogenation reactions as a result of oscillations in temperature gradients within a catalyst particle. This phenomenon occurs due to coupling of heat and mass transfer effects (and phase transition) with chemical reaction. The combination of NMR spectroscopy with MRI has been used to visualize the spatial distribution of the reactant-to-product conversion within an operating reactor.

REACTOR PERFORMANCE MODELS FOR TRICKLE BED REACTORS

The effective rate analysis discussed earlier needs to be combined with the overall reactor models to simulate the performance of trickle bed reactors.

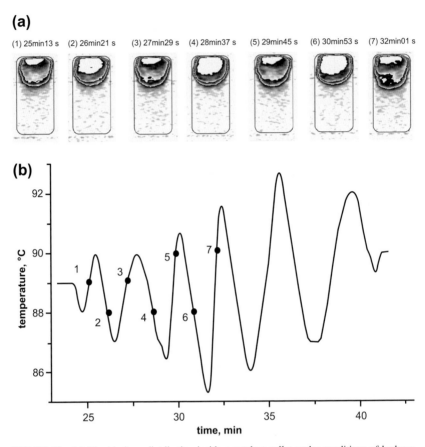

FIGURE 10   (a) Liquid phase distribution inside a catalyst pellet under conditions of hydrogenation of α-methyl styrene and (b) oscillations of temperature in a trickle bed (Lysova et al., 2007).

Many reports on experimental and modeling of trickle bed reactors have been published with the aim of comparing the predictions with experimental data and understanding the interplay between the overall reactor performance and various operating and model parameters (see, for example, Dudukovic et al., 1999; Goto & Smith, 1976). Many of the industrially important reactions involve complex reaction network and may also be highly exothermic. Hence, detailed investigation of such complex multistep reactions and other reactor parameters which influence reactor performance for such reactions is carried out. Most of the trickle bed reactor performance studies reported in the literature are under isothermal conditions and involve simple reaction schemes. Investigation of trickle bed reactor performance for important complex multistep reactions is scarce in the literature (Bergault, Rajashekharam, Chaudhari, Schweich, & Delmas, 1997; Chaudhari et al., 2002; Huang & Kang,

1995; Khadilkar et al., 1998; Rajashekharam et al., 1998). Some recent investigations into fixed-bed reactor performance studies are listed in Table 1.

Most of the studies were carried out under isothermal conditions and have considered pseudo-homogeneous models based on the assumption of plug flow (Satterfield, 1975; Shah, 1979) or heterogeneous models with plug flow of both gas and liquid phases (Jiang et al., 1998; Rajashekharam et al., 1998). Some models accounted for liquid flow non-uniformity and maldistribution by using an axial dispersion model (ADM). Most of the investigations deal with hydrogenation or oxidation reactions in pure or moderately concentrated organic or aqueous solutions (large excess of liquid reactant). Liquid reactants/ solvents were assumed to be non-volatile and gas phase was assumed to be pure at constant partial pressure of the reacting gas. The key effect that was incorporated into most recent models was that of partial wetting and transport of gaseous reactant to externally dry areas of the catalyst resulting in higher rates observed in most experimental data.

It was observed generally that under conditions of gas phase reactant limitation, the rates decreased with increase in liquid velocity due to higher catalyst wetting at higher liquid velocities (and hence reduced direct gas–solid mass transfer). Under conditions of liquid phase reactant limitation, the rates were found to decrease with lower wetting due to decrease in the effective liquid–solid contacting. Al-Dahhan and Dudukovic (1996) achieved an improvement in the performance of trickle bed reactor operating under conditions where limiting reactant is in the liquid phase by the use of fines in the bed. The addition of fines improved the liquid–solid contacting efficiency and a corresponding increase in the rate was observed. Some models considered non-isothermal effects and used pseudo-homogeneous energy balance to estimate the temperature at any axial location. Other variants include a cell model, a cross-flow model, mixing cell model, and some others based on liquid flow maldistribution or stagnant liquid zones in the reactor.

Among the reactor-scale models, the plug flow model has been more commonly used for the modeling of trickle bed reactor. ADM is the simplest model describing differential mixing in fixed-bed reactors by superimposing axial dispersion on plug flow. The ADM involves only one parameter, the axial dispersion coefficient, usually expressed as a Peclet number. The mixing cell model (Ramachandran and Smith, 1979b) considers a flow through a series of mixing cells in the interstices of the packing, where the flow is characterized by the number of cells in series and the liquid holdup. However, this model does not adequately represent the actual flow in a fixed bed but is useful to deal with complex multistep reactions. The cross-flow model has been suggested to account for the considerable liquid stagnancy. It assumes that liquid holdup can be split into two parts: stagnant pockets and dynamic flowing liquid in plug flow, with exchange between the two. The cross-flow model requires three parameters: the fraction of the plug flow, the exchange coefficient, and the external liquid holdup.

TABLE 1 Examples of Trickle Bed Reactor Modeling and Experimental Verification Studies

S. no.	Reaction/Reactor	Rate Analysis	Model Assumptions	Reference
1	H_2O_2 decomposition/trickle bed reactor (TBR)	Linear kinetics	Isothermal, partial wetting, two-region cell model	Sims, Gaskey, and Luss (1994)
2	Hydrogenation of C_4 olefins/UFR*	L–H kinetics	Isothermal, plug flow	Vergel, Euzen, Trambouze, and Wauquier (1995)
3	Hydrogenation of 3-hydroxypropanal/TBR	L–H kinetics	Isothermal, plug flow, partial wetting, heat balance	Valerius et al. (1996)
4	Hydrotreating of vacuum gas oil/TBR	L–H kinetics	Isothermal, plug flow, partial wetting	Korsten and Hoffmann (1996)
5	Hydrogenation of α-methyl styrene/TBR and UFR	L–H kinetics	Isothermal, plug flow, partial wetting, high pressure	Khadilkar et al. (1996)
6	Selective hydrogenation of 1,5,9-cyclododecatriene/UFR	Linear kinetics	Isothermal, axial dispersion, high pressure/temperature	Stuber, Benaissa, and Delmas (1995)
7	SO_2 oxidation/TBR	L–H kinetics	Isothermal, full wetting	Ravindra, Rao, and Rao (1997)
8	SO_2 oxidation/TBR/UFR	L–H kinetics	Isothermal, partial wetting, axial dispersion, static—dynamic	Iliuta and Iliuta (1997)
9	Phenol oxidation/TBR	L–H kinetics	Isothermal, full wetting, plug flow, high pressure/temperature	Pintar, Bercic, and Levec (1997)
10	Hydrogenation of α-methyl styrene/TBR	Linear kinetics	Isothermal, plug flow, partial wetting	Castellari et al. (1997)

11	Hydrogenation of acetophenone/TBR	L–H kinetics	Non-isothermal, plug flow, full wetting, high pressure/temperature	Bergault et al. (1997)
12	Hydrogenation of unsaturated ketones in supercritical CO_2/TBR	Chapter 4; power law kinetics	Non-isothermal, plug flow, full wetting	Devetta, Canu, Bertucco, and Steiner (1997)
13	Hydrogenation of 2,4-dinitrotoluene/TBR	L–H kinetics	Non-isothermal, plug flow, partial wetting, stagnant liquid	Rajashekharam et al. (1998)
14	Hydrogenation of α-nitromethyl-2-furanmethanol/TBR	L–H kinetics	Isothermal, plug flow, partial wetting	Jiang et al. (1998)
15	Oxidation of substituted phenols/TBR	Linear kinetics	Isothermal, plug flow	Tukac and Hanika (1998)
16	Hydrogenation of maleic anhydride/UFR	L–H kinetics	Isothermal, axial dispersion, full wetting	Herrmann and Emig (1998)
17	Hydrogenation of 1,5,9-cyclododecatriene/TBR and UFR	L–H kinetics	Non-isothermal, plug flow, partial wetting, stagnant liquid	Chaudhari et al. (2002)
18	Hydrogenation of 1,5,9-cyclododecatriene/TBR	Eley–Rideal kinetics	Non-isothermal heterogeneous model, partial wetting effect	Dietz, Julcour, and Delmas (2003)
19	Catalytic hydroprocessing of oil feedstock	L–H kinetics	Non-isothermal, homogeneous plug flow axial dispersion model	Avraam and Vasalos (2003)
20	Hydrogenation of α-methyl styrene and phenol oxidation/TBR and UFR	L–H kinetics	Axial dispersion model	Guo and Muthanna (2004)
21	Catalytic wet air oxidation/TBR and UFR	L–H kinetics	Axial dispersion model for liquid phase coupled with cell stack model for gas phase	Guo and Al-Dahaan (2005)

(Continued)

TABLE 1 Examples of Trickle Bed Reactor Modeling and Experimental Verification Studies—Cont'd

S. no.	Reaction/Reactor	Rate Analysis	Model Assumptions	Reference
22	Wet air oxidation of phenol	Power law kinetics	Non-isothermal, plug flow model, partial wetting effect	Suwanprasop et al. (2005)
23	Hydrotreating of benzene/TBR	L–H kinetics	Non-isothermal, one-dimensional and two-dimensional cell network models, radial liquid maldistribution, partial wetting effect	Guo et al. (2008)
24	Hydrogenation of benzene/TBR	L–H kinetics	Non-isothermal heterogeneous three-phase model, Maxwell–Stefan mass transfer model, and effective diffusivity model	Roininen et al. (2009)
25	Catalytic oxidation of phenol/TBR	L–H kinetics	Isothermal, axial dispersion model, plug flow model	Wu et al. (2009)

*Upflow reactor

In some studies, cell network model has been applied for trickle bed reactor modeling, in which, fixed bed was approximated as a cylindrically symmetrical network of perfectly stirred tank reactors. The reactants were envisioned to enter any given stirred tank as a single phase from the two preceding tanks. Alternative rows were offset at half a tank to allow for radial mixing. The effluent from the stirred tank was then fed through subsequent stages. Jaffe (1976) applied this concept to the heat release of a single-phase hydrogenation process and simulated the occurrence of steady-state hotspots due to flow maldistribution. Schnitzlein and Hofmann (1987) developed an alternative cell network model in which the elementary unit consisted of an ideal mixer and a subsequent plug flow unit. These fluid streams were split or merged in infinitesimally small adiabatic mixing cells (without reaction), located between the different layers of the elementary units. Kufner and Hofmann (1990) incorporated the radial porosity distribution into the above cell model, which led to a better agreement of the predicted temperature profile with the experimental data. The cell network models mentioned above were examined for single-phase flow with offset in alternative rows of cells. Recently, Guo, Jiang, and Al-Dahaan (2008) have developed a trickle bed reactor model, which is capable of handling multiphase flow and reactions as well as temperature change due to the phase transition and flow maldistribution. The model serves as a guide to understand the reactor performance and optimization.

Pellet-scale diffusion with reaction was studied by taking reactant limitation into account in simpler versions and in general case, by considering both gas and liquid phase reaction zones and solution of gas—liquid interface by considering liquid inhibition, pore filling, and capillary condensation in a partially internally wetted pellet (Harold & Watson, 1993). Approximate solutions from gas—solid catalyst level equations have also been verified by numerical solution for n-th order as well as Langmuir—Hinshelwood-type kinetics. Some selected studies and observations in trickle bed reactors where performance studies have been carried out are summarized in the sections below.

El-Hisnawi (1981) employed the heterogeneous plug flow model for low pressure operation in a trickle bed reactor to solve the gas-limited reaction problem due to externally dry catalyst areas. An overall wetting efficiency was introduced in the plug flow model to account for the partial wetting of the catalyst particles. The key effect incorporated in such a model was that of partial wetting and transport of gaseous reactant to dry external areas of the catalyst. The direct contact of gas and solid results in higher rates which are also observed in the experiments for gas-limited reactions. The surface concentration of the limiting reactant was obtained by solution of the reaction transport equation at the catalyst surface, and then it was substituted into the plug flow equation to obtain the profile for the non-limiting reactant along the reactor length. Analytical solutions were derived for first-order kinetics for the resulting coupled linear ordinary differential equations at low pressures.

Khadilkar, Wu, Al-Dahhan, and Dudukovic (1996) have employed El-Hisnawi's plug flow model for hydrogenation of α-methyl styrene (AMS) at high pressure (1.5 MPa), where a numerical solution was demanded due to non-linear kinetics exhibited by the reaction. The pellet effectiveness factor in El-Hisnawi model was fitted at one space–time with the experimental observation. Then the effectiveness factor was used as a fitting parameter for the other space–time, which, however, in no way reflected the actual mass transfer resistance inside the pellet and the pellet external wetting with the liquid.

Hydrogenation of benzaldehyde to benzyl alcohol was studied in a slurry reactor using nickel catalyst by Herskowitz (1991) and the Langmuir–Hinshelwood-type of kinetic model was used to predict the performance of a trickle bed reactor for the abovementioned reaction. A completely wetted catalyst model was developed and all the mass transfer resistances were considered in the model assuming that hydrogen is the rate-limiting reactant. The study was carried out for a range of gas velocity, temperature, and pressure. The effectiveness factor was found to be very low indicating the presence of strong intraparticle mass transfer resistance. The gas–liquid mass transfer was found to be important, whereas the liquid–solid mass transfer was found to be negligible. It was concluded that slurry reactor was a better choice than a trickle bed reactor for such rapid reactions.

The impact of partial external wetting and liquid–vapor phase equilibrium on the catalyst performance was investigated for the hydrogenation of naph-thalene over Pt/Al$_2$O$_3$ catalyst in a trickle bed reactor by Huang and Kang (1995) using solvents of varying volatility. When the gas–liquid flow ratio was increased, the wetting efficiency of the catalyst particles decreased, thus enhancing the reaction rates through direct gas–solid mass transfer. In order to understand the effect of solvent volatility on the reactor performance, naph-thalene was dissolved in various solvents like n-hexadecane, n-dodecane, and n-octane and the effect of gas velocity on the reactor performance was studied. It was observed that the rate of hydrogenation increased with increasing hydrogen flow rate. It was shown experimentally as well as theoretically that the lower the boiling point of the solvent more the extent of solvent evaporation with increasing gas flow rate.

In an effort to represent the performance of a trickle bed reactor for the hydrogenation of 3-hydroxypropanal to 1,3-propanediol, Valerius, Zhu, Hofmann, Arntz, and Haas (1996) considered two different approximations of the overall catalyst effectiveness factor: (1) the effectiveness factors of dry, half-wetted and totally wetted slabs were evaluated using a model proposed by Beaudry, Duducovic, and Mills (1987) and (2) a new cylinder shell model was used, leading to one-dimensional mass balance equations inside the porous catalyst particle for all possible values of the external wetting efficiency on the particle scale. The aim of the investigation was to study the role of mass transfer and degree of wetting on the trickle bed reactor performance. It was found that, both the abovementioned parameters affect the reaction rates. For both the models, parametric values of $f_w = 0.6-1.0$ led to a higher pressure dependence and on increasing the liquid velocity, a maximum in the reaction rate was observed. A reactor model for hydrotreating reactions in a pilot-scale trickle bed reactor was developed by Korsten and Hoffmann (1996). The model, based on the two-film theory, was tested with regard to hydrodesulfurization of vacuum gas oil in a high pressure pilot plant reactor under isothermal conditions. The axial dispersion in both phases was found to be negligible and various mass transfer coefficients, gas solubility, and other properties of the gas as well as the liquid phases were determined using various correlations. The kinetics of the reaction was represented by a Langmuir—Hinshelwood model and the intraparticle mass transfer within the catalyst pellets represented through catalyst effectiveness factor. The sulfur content of the product oil was found to depend strongly on the gas/oil flow ratio within the reactor. The poor conversion observed was explained as a result of incomplete catalyst wetting properties. The simulation showed good agreement with the experimental values for a wide range of temperature, pressure, space velocity, and gas/oil ratio.

Castellari, Cechini, Gabarian, and Haure (1997) studied the hydrogenation of AMS to cumene over Pd/Al_2O_3 in a laboratory trickle bed reactor operated at low liquid velocities. Under these conditions, the catalyst pellets were incompletely wetted and also vaporization of liquid phase reactants was observed. The presence of gas—solid catalysis enhanced the reaction rate significantly. To improve the mass transfer between the phases, the length of inert post-packing section was increased and the inert phase acted as an absorber of cumene produced in gas phase. Experimental global rates determined from the liquid cumene concentrations varied with the post-packing length. An one-dimensional model taking into consideration three different zones, first pre-packing inert zone where liquid feed containing AMS gets saturated with H_2, catalytic bed, and post-packing inert section acting as an absorber of the gas phase cumene produced, was found to agree well with the observed experimental results.

Liquid phase hydrogenation of acetophenone using Rh/C catalyst was studied by Bergault et al. (1997) in trickle bed reactor and slurry airlift reactor and the performances of these two reactors were compared in terms of productivities and yields. For modeling trickle bed reactor performance, a non-isothermal plug flow reactor model incorporating the external and intraparticle mass transfer effects was developed. It was assumed that the catalyst was fully wetted and the mass and heat transfer correlations were estimated using various correlations available in the literature. It was concluded that the available correlations for gas—liquid mass transfer are not satisfactory and catalyst wetting is an important process that cannot be neglected in the modeling and intraparticle diffusion effects played an important role in determining the reactor efficiency.

Rajashekharam et al. (1998) reported the experimental verification of a non-isothermal trickle bed reactor model for the hydrogenation of 2,4-dinitrotoluene using 5%Pd/Al$_2$O$_3$ catalyst, incorporating the partial wetting of the catalyst as well as the stagnant liquid holdup. It was assumed that the catalyst can be divided into three zones, where the catalyst is exposed to dynamic liquid flow, static liquid pockets, and gas phase. The external mass transfer effects and intraparticle diffusion effects of the gaseous reactant hydrogen were taken into consideration in the model and a Langmuir–Hinshelwood type of rate model was used to describe the intrinsic kinetics of the reaction. The effect of various parameters like liquid velocity, gas velocity, and temperature and catalyst particle size on rate of hydrogenation, conversion, and temperature rise inside the reactor was investigated. It was found that on increasing the liquid velocity, even though the conversion became low, the global rate of hydrogenation increased. The effect of gas velocity on the reactor performance was found to be negligible. The sensitivity of the model to gas–solid mass transfer, gas–liquid mass transfer, liquid–solid mass transfer, and the role of stagnant liquid pockets was also investigated. It was found that the importance of parameters in the descending order was gas–solid > gas–liquid > liquid–solid and also concluded that the contribution by stagnant liquid pockets was negligible.

Another detailed study of a complex hydrogenation in a trickle bed reactor is the manufacture of a herbicide intermediate α-aminomethyl-2-furanmethanol from α-nitromethyl-2-furanmethanol using Raney nickel catalysts presented in two parts by Jiang et al. (1998) and Khadilkar et al. (1998). The first part deals with the experimental investigation of the effect of operating parameters on yields and productivity in a trickle bed reactor. It was found that trickle flow pattern of liquid was preferable in getting the desired product, due to high ratio of catalyst to liquid volumes. For a given catalyst and under a particular operating pressure, the yield of the desired product improved by decreasing the feed concentration of substrate and by decreasing the liquid flow rate. Lower temperatures also lead to improvement in desired product selectivity and diluting the bed with solid fine particles resulted in a better utilization of the catalyst. In the second part, a kinetic scheme for the complex reaction was developed and based on a trickle bed reactor model and parameter optimization programs, apparent kinetic constants were obtained using the experimental data.

Enhancing the reactor performance by using various techniques has been of great research interest in recent years. The periodic operation of liquid flow in trickle bed reactor has shown to enhance the reaction rate in the case of a gas-limited reaction. The liquid flow on switching off results in draining of the catalyst bed and thus reducing the mass transfer resistance offered by the liquid phase and facilitates the direct mass transfer from the gas phase to solid phase. The effect of such a periodic flow reversal for the oxidation of SO$_2$ under conditions of gas-limiting reactant was studied by Haure, Hudgins, and Silveston (1989) and found an increase in reaction rate in the range of 30–40%. If the reaction is exothermic, such a break in liquid flow results in poor heat dissipation resulting in a temperature rise in bed thus further enhancing the reaction rate. By exploiting the exothermicity of the reaction, Castellari and Haure (1995) were able to get a drastic increase of about 400% in rate for hydrogenation for AMS over palladium catalyst. Formation of hotspots and evaporation of liquid phase leading to partial gas phase reaction were responsible for such a rate enhancement. Al-Dahhan and Dudukovic (1996) achieved an improvement in the performance of trickle bed reactor operating under conditions where limiting reactant was in the liquid phase by using the fines. The addition of fines improved the liquid–solid contacting efficiency and a corresponding increase in the rate was observed.

Iliuta and Larachi (2001) have discussed the design of reactors by exploring a parallel modeling framework in which the reactor and particle scales were considered. The reactor-scale piston dispersion exchange (PDE) model was employed to capture both transient and space dependences of reactants in the dynamic and static liquid zones. The mass transfer processes between gas–liquid, gas–solid, and dynamic liquid–static liquid were integrated. Chemical reaction, as a sink to deplete the reactants, was coupled with the PDE model. At the pellet scale, they solved the general diffusion reaction equation to supply the time and space distribution of the reactant concentrations. They employed a parallel solution strategy to simultaneously resolve the pellet and the reactor-scale model for the catalytic wet air oxidation of phenol in fixed bed with MnO_2/CeO_2 catalyst pellets.

Guo et al. (2008) have developed one-dimensional and two-dimensional cell network models to simulate steady-state behavior of the trickle bed reactor employed for highly exothermic hydrotreating of benzene. A comprehensive one-dimensional mixing cell model was developed to account for the phase transition in pilot plant reactors devoid of radial flow maldistribution. This model includes the local changes of phase velocities, species concentrations, external wetting efficiency, liquid holdup, and mass transfer rate due to phase transition and their effects on the reaction rates. The one-dimensional model was applied in the reaction system of benzene hydrogenation to cyclohexane in order to predict the temperature profiles and the change of species concentrations along the reactor axis. After the one-dimensional model was validated against the experimental temperature profile reported in the literature, it was extended to a two-dimensional model to assess the impact of flow maldistribution on the formation of hotspots. The two-dimensional model was able to take advantage of the validated one-dimensional model and a new solution scheme was designed to expedite the solution process and to enhance the solution stability. The ultimate goal was to develop a reliable model that will enable a rational design and control to avoid the undesirable formation of hotspots.

Roininen, Alopaeus, Toppinen, and Aitamma (2009) have developed a heterogeneous model to simulate the industrial trickle bed reactor for benzene hydrogenation and simulated temperature profiles were compared with the actual plant data. Analysis of the simulation results showed that the process is limited by hydrogen mass transfer through the gas—liquid interface. Therefore, the mass transfer correlations, especially for $k_L a$, must be carefully chosen. A completely wetted catalyst model was also developed and all the mass transfer resistances were considered in the model assuming that hydrogen is the rate-limiting reactant. The study was carried out for a varying range of gas velocity, temperature, and pressure. The effectiveness factor was found to be very low indicating the presence of strong intraparticle mass transfer resistance. The gas—liquid mass transfer was found to be important, whereas the liquid—solid mass transfer was found to be negligible. Some important models developed for trickle bed reactors are briefly discussed in the following.

Empirical Pseudo-Homogeneous Models

Several empirical models were proposed in early stages of development to describe the conversion in a trickle bed reactor in which the rate was assumed to be proportional to the wetted fraction of the catalyst particles and liquid reactant concentrations under conditions of constant gas phase concentration (applicable for most hydrogenation reactions and hydroprocessing technologies in petroleum industry). These models were essentially developed to describe empirically the influence of operating variables on conversion. The various empirical models proposed are summarized below (from Hofmann, 1986):

Plug flow model

$$-\ln \frac{C}{C_f} = (1 - \varepsilon)\eta_0 k_v (L/u_l) = (1 - \varepsilon)\eta k_v 3600 (\text{LHSV})^{-1} \quad (26)$$

Axial dispersion model (Wehner & Wilhelm, 1956):

$$-\ln \frac{C}{C_f} = (1 - \varepsilon)\eta_0 k_v 3600 (\text{LHSV})^{-1} - \frac{1}{Bo}(1 - \varepsilon)^2 \eta_0^2 k_v^2 3600^2 (\text{LHSV})^{-2}$$

$$(27)$$

External holdup model (Henry & Gilbert, 1973):

$$-\ln \frac{C}{C_f} \propto (1 - \varepsilon)\eta_0 k_v 3600 (\text{LHSV})^{-0.66} L^{0.33} d_p^{-0.66} v_t^{0.33} \quad (28)$$

Effective wetting model (Mears, 1974):

$$-\ln \frac{C}{C_f} \propto (1 - \varepsilon)\eta_0 k_v 3600 (\text{LHSV})^{-0.68} L^{0.32} d_p^{0.18} v_t^{-0.05} (G_c/G_w)^{0.21} \quad (29)$$

While these models are useful as a first approximation, they do not provide insight into the contribution of specific reaction kinetics, mass and heat transfer and mixing effects, and particularly are unsuitable for complex multistep reactions. More rigorous models are required to understand the trickle bed reactors and engineering complexities associated with it.

Generalized Model for Complete Wetting of Catalyst Particle

Trickle bed reactor performance depends on several phenomena that operate in parallel or sequential. A careful account of these aspects is essential for complete understanding of their influence on overall performance in terms of conversion, selectivity, and temperature profiles. If we consider a bimolecular reaction of the type: $A(g) + bB(l) \rightarrow$ products, following steps will occur,

(a) transport of gas phase reactant A from gas to liquid phase
(b) transport of dissolved A and liquid phase reactant B from the liquid phase to surface of the catalyst
(c) intraparticle diffusion of A and B within the catalyst pores
(d) simultaneous chemical reaction on the catalyst surface, and
(e) desorption of products

In addition, intraparticle heat and mass transfer steps are also required to be accounted for. A general form of mathematical model that describes the variation of concentrations of A and B in a trickle bed reactor is given below for a bimolecular reaction under non-isothermal conditions:

Gas phase balance:

$$D_{eg} \frac{d^2 A_g}{dx^2} - u_g \frac{dA_g}{dx} = k_L a_B \left(\frac{A_g}{H_A} - A_l \right) \tag{30}$$

Liquid phase balance:

$$D_{el} \frac{d^2 A_l}{dx^2} - u_l \frac{dA_l}{dx} + k_L a_B \left(\frac{A_g}{H_A} - A_l \right) = (k_s a_p)_A (A_l - A_s)$$

$$= (1 - \varepsilon_B) \eta_c k_2 A_s B_s \tag{31}$$

$$D_{el} \frac{d^2 B_l}{dx^2} - u_l \frac{dB_l}{dx} = (k_s a_p)_B (B_l - B_s) = (1 - \varepsilon_B) \eta_c k_2 A_s B_s \tag{32}$$

Heat balance:

$$-\lambda_e \frac{d^2 T}{dx^2} + (u_l \rho_l C_{pl} + u_g \rho_g C_{pg}) \frac{dT}{dx} = (1 - \varepsilon_B)(-\Delta H_R) \eta_c k_2 A_s B_s \tag{33}$$

Boundary conditions:

$$x = 0, \quad D_{eg} \frac{dA_g}{dx} = u_g (A_g - A_{gi})$$

$$D_{el}\frac{dA_1}{dx} = u_1(A_1 - A_{1i}) \text{ and } \lambda_e\frac{dT}{dx} = h_r(T - T_i) \tag{34}$$

$$x = L, \quad \frac{dA_g}{dx} = \frac{dA_1}{dx} = \frac{dT}{dx} = 0 \tag{35}$$

This model considers overall heat effects and assumes that temperature in all phases at any axial position is uniform but varies along the length of the reactor and described in terms of overall heat transfer coefficient. Intraparticle heat transfer is neglected. The above equations can be easily extended for any other reaction kinetics.

Ramachandran and Chaudhari (1983) developed models for non-linear kinetics under isothermal conditions and also for special cases with gas-limiting or liquid-limiting reactions.

Adiabatic Trickle Bed Reactor Model

In many situations, the trickle bed reactor is operated under adiabatic conditions, for which a reactor performance model has been proposed (Shah & Paraskos, 1975; Shah, Stuart, & Sheth, 1976) with the following assumptions: (a) the catalyst is completely wetted, (b) evaporation of liquid phase components is negligible, (c) the limiting reactant is present in liquid phase, the gas phase concentration being uniform throughout the reactor, (d) the radial temperature and concentration variation is negligible, and (e) external mass and heat transport resistances are unimportant (i.e., pseudo-homogeneous model). The model equations are as follows.

$$-u_1\frac{dB_1}{dx} = wk_n(T)B_1^n \tag{36}$$

where $k_n(T) = k_{mn}(A^*)^m$ and is a function of temperature. The heat balance of the reactor is,

$$(u_g C_{pg}\rho_g + u_1 C_{pl}\rho_L)\frac{dT}{dx} = (-\Delta H)wk_n(T)B_1^n \tag{37}$$

$$k_n(T) = k_n(T_i)\exp\left[\frac{E_0}{R_g T_i}\left(1 - \frac{T_i}{T}\right)\right] \tag{38}$$

Eliminating the kinetic term in the above equations, we get

$$\theta = \frac{T}{T_i} = 1 + \beta_H - \beta_H b_1 \tag{39}$$

$$-\frac{db_1}{dz} = \alpha_{ro}b_1^n \exp\left[Ar\left(1 - \frac{1}{1 + \beta_H - \beta_H b_1}\right)\right] \tag{40}$$

$$\beta_H = \frac{(-\Delta H)B_{li}}{T_i C_{pl}\rho_L[1 + (u_g C_{pg}\rho_G/u_i C_{pl}\rho_L)]} \quad (41)$$

$$Ar = \frac{E_a}{R_g T_i} \quad (42)$$

$$\alpha_{ro} = \frac{k_n(T_i)B_{li}^{n-1}L}{u_1} \quad (43)$$

Initial conditions:

$$z = 0, \ b_1 = 1, \text{ and } \theta = 1 \quad (44)$$

These model equations can be solved using the standard numerical methods, used for solving ordinary differential equations, to predict the conversion and temperature profiles along a trickle bed reactor.

Non-isothermal Trickle Bed Reactor Model: Complex Reactions

A detailed analysis of a trickle bed reactor model incorporating the influence of partial wetting, stagnant liquid pockets, external as well as intraparticle mass transfer limitations has been developed by Rajashekharam et al. (1998) and then by Chaudhari et al. (2002) for complex reactions schemes involved in catalytic hydrogenations. An example representing hydrogenation of cyclododecatriene is described here as an illustration:

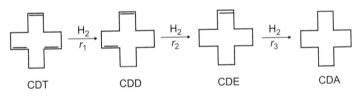

In this model, the following steps have been considered.

(i) A spherical catalyst particle is assumed to be divided into three zones that represented (a) a dry unwetted zone, (b) a wetted zone covered by the flowing dynamic liquid, and (c) a wetted zone covered by the stagnant liquid as shown in Figure 7.

(ii) The gas and liquid phases are assumed to be in plug flow with a non-volatile liquid phase reactant in excess compared to the gaseous reactant concentration (e.g., hydrogen or any sparingly soluble gas),

(iii) Gas–liquid, liquid–solid, and intraparticle mass transfer resistances for the gas phase reactant are considered whereas the mass transfer resistances for the liquid components are assumed to be negligible,

(iv) Intraparticle and interparticle heat transfer resistances are negligible but the overall heat transfer (bed to wall) is considered to incorporate the non-isothermal effects,

(v) The overall catalytic effectiveness factor is assumed to be a sum of weighted average of the effectiveness factors in the dynamic liquid covered, stagnant liquid covered, and unwetted gas-covered zones, respectively, represented by Eq. (24), and

(vi) The catalyst particle is assumed to be completely wetted internally. In Eq. (24), η is the overall catalytic effectiveness factor and f_d and f_s are the fractions covered by dynamic flowing and stagnant liquids. η_d, η_s and η_g are the overall catalytic effectiveness factors for dynamic, stagnant, and dry zones, respectively.

For a reaction scheme of the type:

$$A(g) + B(l) \rightarrow C(l)$$
$$A(g) + C(l) \rightarrow E(l)$$
$$A(g) + E(l) \rightarrow P(l) \tag{46}$$

The equations for overall rate of reaction of A were developed following Ramachandran and Chaudhari (1983) and Bischoff (1965) for conditions of significant external and intraparticle mass transfer gradients. The following reaction kinetics was considered:

$$r_1 = \frac{wk_1 BA^*}{(1 + K_B B + K_C C + K_E E)} \tag{47}$$

$$r_2 = \frac{wk_2 CA^*}{(1 + K_B B + K_C C + K_E E)} \tag{48}$$

$$r_3 = \frac{wk_3 EA^*}{(1 + K_B B + K_C C + K_E E)} \tag{49}$$

Under the conditions of significant intraparticle gradients for the gas phase reactant A and when the liquid phase reactant is in excess, the overall rate of hydrogenation can be expressed as:

$$R_A = \frac{\eta_0 w(k_1 B_l + k_2 C_l + k_3 E_l)A^*}{(1 + K_B B_l + K_C C_l + K_E E_l)} \tag{50}$$

where, η_0 is given by Eq. (24). For conditions of complete wetting and absence of stagnant liquid pockets, effectiveness factor becomes equal to η_c which is given in Eq. (51) for the spherical catalyst particles:

$$\eta_c = \frac{1}{\phi}\left(\coth 3\phi - \frac{1}{3\phi}\right), \tag{51}$$

With

$$\phi = \frac{R}{3}\left(\frac{S_p(k_1 B_l + k_2 C_l + k_3 E_l)}{D_e(1 + K_B B_l + K_C C_l + K_E E_l)}\right)^{1/2}, \tag{52}$$

Or in dimensionless form:

$$\phi = \phi_0 \left(\frac{(b_1 + k_{21}c_1 + k_{31}e_1)}{(1 + k_b b_1 + k_c c_1 + k_e e_1)} \right)^{1/2}, \tag{53}$$

With

$$\phi_0 = \frac{R}{3} \left(\frac{\rho_p k_1 B_{li}}{D_e} \right)^{1/2}, \tag{54}$$

The dimensionless forms of the mass balance equations for A, B, C and E are

$$\frac{da_{ld}}{dz} + \alpha_{gl}(1 - a_{ld}) = \frac{\eta_c \alpha_r (b_1 + k_{21}c_1 + k_{31}e_1)}{(1 + k_b b_1 + k_c c_1 + k_e e_1)} \times \left\{ \frac{f_d a_{ld}}{(1 + \eta_c \phi^2 / N_d)} \right.$$
$$\left. + \frac{f_s a_{ld}}{(1 + \eta_c \phi^2 / N_s) + (\eta_c \phi^2 / \alpha_s N_s)} \right\} \tag{55}$$

with

$$(1 - f_d - f_s)\alpha_{gs}(1 - a_{sg}) = (1 - f_d - f_s) \times \frac{\eta_c \alpha_r (b_1 + k_{21}c_1 + k_{31}e_1)}{(1 + k_b b_1 + k_c c_1 + k_e e_1)}$$
$$\times \left[\frac{1}{(1 + \eta_c \phi^2 / N_g)} \right] \tag{56}$$

$$-\frac{db_{ld}}{dz} = \frac{\eta_c \alpha_r b_1 \chi}{q_B (1 + k_b b_1 + k_c c_1 + k_e e_1)} \tag{57}$$

$$\frac{dc_{ld}}{dz} = \frac{\eta_c \alpha_r (b_1 - k_{31}c_1)\chi}{q_B (1 + k_b b_1 + k_c c_1 + k_e e_1)} \tag{58}$$

$$\frac{de_{ld}}{dz} = \frac{\eta_c \alpha_r (k_{21}c_1 - k_{41}e_1)\chi}{q_B (1 + k_b b_1 + k_c c_1 + k_e e_1)} \tag{59}$$

$$\frac{dP_{ld}}{dz} = \frac{\eta_c \alpha_r k_{31}e_1 \chi}{q_B (1 + k_b b_1 + k_c c_1 + k_e e_1)} \tag{60}$$

with

$$\chi = \frac{f_d a_{ld}}{(1 + \eta_c \phi^2 / N_d)} + \frac{f_s a_{ld}}{[(1 + \eta_c \phi^2 / N_s) + (\eta_c \phi^2 / N_s)]} + \frac{(1 - f_d - f_s)}{(1 + \eta_c \phi^2 / N_g)} \tag{61}$$

The heat evolved during the reaction was assumed to be carried away by the flowing liquid and transfer to the reactor wall, which is characterized by the bed-to-wall heat transfer coefficient, U_w. Under such conditions, where the interphase and intraparticle heat transfer resistances are assumed to be negligible, the heat balance of the reactor can be expressed in dimensionless form as:

$$\frac{d\theta}{dz} = \frac{\eta_c \alpha_r (b_1 + k_{21}c_1 + k_{31}d_1)\chi}{q_B(1 + k_B b_L + k_C c_1 + k_E e_1)} - \beta_2(\theta_b - \theta_W) \tag{62}$$

The initial conditions are

$$z = 0; \quad a_1 = b_1 = 1; \quad c_1 = e_1 = p_1 = 0; \quad \text{and} \quad \theta = 1 \tag{63}$$

The solution of the above model equations allows us to predict the conversion of reactant B and the overall rate of reaction, R_A given as:

$$X_B = 1 - b_1 \tag{64}$$

and

$$R_A = \frac{U_1}{L}(C_1 + 2E_1 + 3P_1) \tag{65}$$

The dimensionless model parameters used in this analysis are presented in Table 2.

Chaudhari et al. (2002) used this model to illustrate an application for hydrogenation of CDT, and some of the typical results comparing the experimental data with theoretical predictions are shown in Figs. 11–14.

Periodic Operations in Trickle Bed Reactors

For exothermic reactions in a trickle bed, the introduction of a reactant feed in a periodic manner or using pulsing and stop flow techniques have been shown to be advantageous to enhance the overall rate of reaction on a reactor scale as well as to control the temperature and runaway in the operating process. Some examples are as follows:

(1) Haure et al. (1989) showed that using on–off flow of liquid in oxidation of SO_2 over activated C catalyst leads to a significant increase in oxidation rates. The rate enhancement is attributed to a higher bed temperature. It is proposed that the on–off liquid flow causes a rising temperature front to move in the bed followed by a falling front in the direction of the liquid flow. A dynamic trickle bed reactor model incorporating gas–liquid mass transfer, product inhibition of reaction, and evaporation has been proposed which predicted the observed moving waves in a trickle bed.

(2) Yamada and Goto (1997) showed that a periodic introduction of water in a trickle bed reactor with four-phase reaction (gas–liquid–liquid–solid) allowed washing and regeneration of the deactivated catalyst.

(3) An experimental studies by Brzić, Schubertb, Häring, Lange, and Petkovsk (2010) and Lange, Gutsche, and Hanika (1999) demonstrated that by changing the liquid flow periodically to force an unsteady state operation leads to increase in time–average conversion in a trickle bed reactor for hydrogenation of AMS.

TABLE 2 Dimensionless Parameters Used for Trickle Bed Reactor Model (Chaudhari, 2002)

Mass transfer parameters	
Gas–liquid mass transfer	$\alpha_g = k_l a_B L / U_l$
Liquid–solid mass transfer	$\alpha_{ls} = k_s a_p L / U_l$
Gas–solid mass transfer	$\alpha_{gs} = k_{gs} a_p L / U_l$
Nusselt number in dynamic zone	$N_d = R k_s / 3 D_e$
Nusselt number in stagnant zone	$N_s = R k_s / 3 D_e$
Nusselt number in dry zone	$N_g = R k_g / 3 D_e$
Thiele parameter	$\phi = R/3 (\rho_p k_l B_{l_i}/D_e)^{0.5} \left[\dfrac{b_l + k_{21} c_l + k_{31} e_l}{1 + k_b b_l + k_c c_l + k_e e_l} \right]$
Exchange parameter for stagnant zone	$\alpha_s = (k_{ex} \varepsilon_l / f_s k_{s,} a_p)$
Heat transfer parameter	
Thermicity parameter	$\beta_1 = \dfrac{(-\Delta H) B_{l_i}}{T_o C_{p_l} \rho_l (1 + [U_g C_{p_g} \rho_g / U_l C_{p_l} \rho_l])}$
Bed-to-wall heat transfer	$\beta_2 = \dfrac{4 U_w L}{d_T C_{p_l} \rho_l (1 + [U_g C_{p_g} \rho_g / U_l C_{p_l} \rho_l])}$
Reaction rate and equilibrium constants	
	$\alpha_r = L w k_l B_{l_i} / U_l$
	$k_{21} = k_2/k_1 ; k_{31} = k_3/k_1$
	$k_b = K_B B_{l_i} ; k_c = K_C B_{l_i}$

(4) The pulsed flow operation of a trickle bed reactor has potential advantages in eliminating the hotspots through the enhanced mass and heat transfer rates. A disadvantage of naturally occurring pulsing flow is the necessity of relatively high gas and liquid flow rates, especially at elevated pressures, resulting in rather short contact times between the phases. These drawbacks of pulsing flow can be eliminated by expanding the pulsing flow regime achieved by periodic operation of a trickle bed (Boelhouwer, Piepers, & Drinkenburg, 1999).

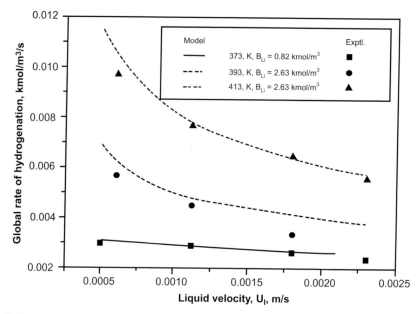

FIGURE 11 Effect of liquid velocity on global rate of hydrogenation in a trickle bed reactor (Chaudhari et al., 2002).

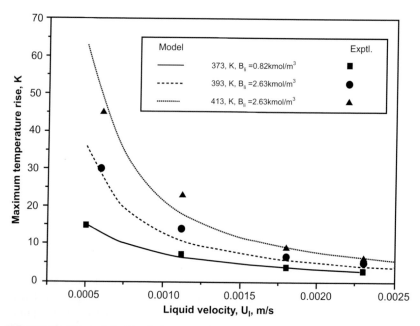

FIGURE 12 Effect of liquid velocity on maximum temperature rise in a trickle bed reactor (Chaudhari et al., 2002).

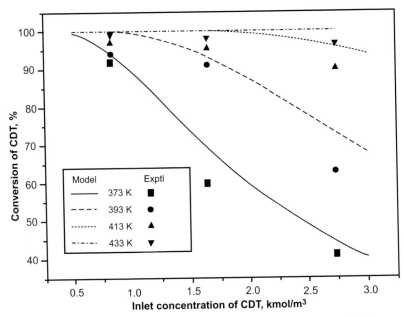

FIGURE 13 Effect of inlet CDT concentration on conversion of CDT in a trickle bed reactor (Chaudhari et al., 2002).

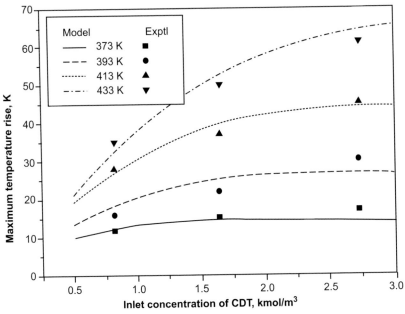

FIGURE 14 Effect of inlet CDT concentration on maximum temperature rise in a trickle bed reactor (Chaudhari et al., 2002).

These studies showed that periodic operation with respect to flow of liquid or reactants can be an important tool not only to control the trickle bed reactor operation but also to give higher productivity. This phenomenon has to be analyzed for specific cases, as a generalization of this model is not practical. More investigations for industrially relevant processes and complex multistep reactions in this area are required.

SUMMARY

The fundamental issues relating to understanding of the rate processes and performance of trickle bed reactors have been discussed in this chapter, which can be useful to interpret the experimental data in laboratory or pilot plant reactors as also the simulation of industrial-scale reactors. The mathematical models accounting for the contributions of reaction kinetics, external and intraparticle mass transfers, and axial and radial mixing for simple and complex reaction kinetics have been summarized. These require several design parameters, which can be predicted using the correlations discussed in Chapter 2. For more detailed understanding of complexities of hydrodynamics and flow behavior, the models presented here can be extended with CFD models discussed in Chapters 4 and 5. Thus, all the elements of developing reactor performance models that would capture relevant issues of scale-up and design are provided in this chapter. It is important to note, however, that while the basic rate and performance models are useful for quantitative analysis of rate-controlling steps for a given process, several issues concerning differences in hydrodynamic and flow behavior at different scales (pilot and industrial) are not easy to assess from these models. A careful judgment of these factors using the approaches discussed in Chapters 4 and 5 needs to be used to develop more general models which would not depend on the scale of operation. It is obvious that such models in generalized forms are not possible and will need to be developed for specific processes.

REFERENCES

Akgerman, A., Collins, G. M., & Hook, B. D. (1985). Effect of feed volatility on conversion in trickle-bed reactors. *Industrial and Engineering Chemistry Fundamentals, 24*, 398–401.

Al-Dahhan, M. H., & Dudukovic, M. P. (1996). Catalyst bed dilution for improving catalyst wetting in laboratory trickle bed reactors. *AIChE Journal, 42*(9), 2595.

Avraam, D. G., & Vasalos, I. A. (2003). HdPro: a mathematical model of trickle bed reactors for the catalytic hydroprocessing of oil feedstocks. *Catalysis Today, 79–80*, 275.

Beaudry, E. G., Duducovic, M. P., & Mills, P. L. (1987). Trickle bed reactors: liquid diffusional effects in a gas limited reaction. *AIChE Journal, 33*, 1435.

Beaudry, E. G., Mills, P. L., & Dudukovic, M. P. (1986). Paper presented at the World congress in Chemical Engineering. Tokyo.

Bergault, I., Rajashekharam, M. V., Chaudhari, R. V., Schweich, D., & Delmas, H. (1997). Modeling of comparison of acetophenone hydrogenation in trickle-bed and slurry airlift reactors. *Chemical Engineering Science, 52*(21/22), 4033.

Bischoff, K. B. (1965). Effectiveness factors for general reaction rate forms. *AIChE Journal, 11*, 351–355.

Boelhouwer, J. G., Piepers, H. W., & Drinkenburg, A. A. H. (1999). Enlargement of the pulsing flow regime by periodic operation of a trickle-bed reactor. *Chemical Engineering Science, 54*, 4661–4667.

Boelhouwer, J. G., Piepers, H. W., & Drinkenburg, A. A. H. (2001). Particle–liquid heat transfers in trickle-bed reactors. *Chemical Engineering Science, 56*, 1181–1187.

Brzić, D., Schubertb, M., Häring, H., Lange, R., & Petkovsk, M. (2010). Evaluation of periodic operation of a trickle-bed reactor based on empirical modeling. *Chemical Engineering Science, 65*(14), 4160–4165.

Castellari, A. T., Cechini, J. O., Gabarian, L. J., & Haure, P. M. (1997). Gas-phase reaction in a trickle-bed reactor operated at low liquid flow rates. *AIChE Journal, 43*(7), 1813.

Castellari, A. T., & Haure, P. M. (1995). Experimental study of the periodic operation of a trickle bed reactor. *AIChE Journal, 41*(6), 1593–1597.

Chaudhari, R. V., Jaganathan, R., Mathew, S. P., Julcour, C., & Delmas, H. (2002). Hydrogenation of 1,5,9-cyclododecatriene in fixed-bed reactors: down- vs. upflow modes. *AIChE Journal, 48*(1), 110–125.

Chaudhari, R. V., & Ramachandran, P. A. (1980). Three phase slurry reactors. *AIChE Journal, 26*, 177–201.

Devetta, L., Canu, P., Bertucco, A., & Steiner, K. (1997). Modeling of a trickle-bed reactor for a catalytic hydrogenation in supercritical CO_2. *Chemical Engineering Science, 52*(21/22), 4163.

Dietz, A., Julcour, C., & Delmas, W. H. (2003). Selective hydrogenation in trickle bed reactor: experimental and modeling including partial wetting. *Catalysis Today, 79–80*, 293.

Dudukovic, M. P., Larachi, F., & Mills, P. L. (1999). Multiphase reactors – revisited. *Chemical Engineering Science, 54*, 1975.

El-Hisnawi, A. A (1981). *Trace and Reaction Studies in Trickle-Bed Reactors. D.Sc. Dissertation.* St. Louis, MS: Washington University.

Germain, A. H., Lefebvre, A. G., L'Homme, G. A. (1974). Experimental study of a catalytic trickle-bed reactor. In, *Chemical Reaction Engineering* II, Advances in Chemistry Series No. 133, ACS, p. 164 (Chapter 13).

Gianetto, A., & Silveston, P. L. (1986). *Multiphase chemical reactors*. Washington, DC: Hemisphere.

Goto, S., Lakota, A., & Levec, L. (1981). Effectiveness factors on nth order kinetics on trickle-bed reactors. *Chemical Engineering Science, 36*, 157–162.

Guo, J., & Al-Dahaan, M. (2005). Modeling catalytic trickle bed and upflow packed bed reactors for wet air oxidation of phenol with phase change. *Industrial and Engineering Chemistry Research, 44*, 6634.

Guo, J., Jiang, Y., & Al-Dahaan, M. H. (2008). Modeling of trickle bed reactors with exothermic reactions using cell network approach. *Chemical Engineering Science, 63*, 751.

Guo, J., & Muthanna, J. G. (2004). A sequential approach to modeling catalytic reactions in packed bed reactors. *Chemical Engineering Science, 59*, 2023.

Hanika, J., Sporka, K., Ruzicka, V., & Rstka, J. H. (1976). Measurement of axial temperature profiles in an adiabatic trickle bed reactor. *Chemical Engineering Journal, 12*, 193–205.

Harold, M. P., & Watson, P. C. (1993). Bimolecular exothermic reaction with vaporization in the half-wetted slab catalyst. *Chemical Engineering Science, 48*(5), 981−1004.

Haure, M., Hudgins, R. R., & Silveston, P. L. (1989). Periodic operation of a trickle-bed reactor. *AIChE Journal, 35*(9), 1437−1444.

Henry, H. C., & Gilbert, J. B. (1973). Scale up of pilot plant data for catalytic hydroprocessing. *Industrial and Engineering Chemistry Process Design and Development, 12*(3), 328−343.

Herrmann, U., & Emig, G. (1998). Liquid phase hydrogenation of maleic anhydride to 1,4-butanediol in a packed bubble column reactor. *Industrial and Engineering Chemistry Research., 37*, 759.

Herskowitz, M. (1991). Hydrogenation of benzaldehyde to benzyl alcohol in a slurry and fixed-bed reactor. *Studies in Surface Science and Catalysis, 59*, 105, Heterogeneous catalysis and fine chemicals -II.

Herskowitz, Carbonell, R. G., & Smith, J. M. (1979). Effectiveness factor and mass transfer in trickle-bed reactors. *AIChE Journal, 25*, 272.

Herskowitz, M., & Mosseri, S. (1983). Global rates of reaction in trickle-bed reactors: effects of gas−liquid flow rates. *Industrial and Engineering Chemistry Fundamentals, 22*, 4.

Huang, T. C., & Kang, B. C. (1995). Naphthalene hydrogenation over Pt/Al$_2$O$_3$ catalyst in a trickle bed reactor. *Industrial and Engineering Chemistry Research, 34*, 2349.

Iliuta, I., & Iliuta, M. C. (1997). Comparison of two-phase upflow and downflow fixed bed reactors performance: catalytic SO$_2$ oxidation. *Chemical Engineering Technology, 20*, 455.

Iliuta, I., & Larachi, F. (2001). Wet air oxidation solid catalysis analysis of fixed and sparged three-phase reactors. *Chemical Engineering and Processing: Process Intensification., 40*, 175.

Jaffe, S. B. (1976). Hot spot simulation in commercial hydrogenation processes. *Industrial and Engineering Chemistry Process Design and Development, 15*(3), 410−416.

Jiang, Y., Khadilkar, M. R., Al-Dahhan, M., Duduković, M. P., Chou, S. K., & Ahmed, G. (1998). Investigation of a complex reaction network: II. Kinetics, mechanism and parameter estimation. *AIChE Journal, 44*(4), 921−926.

Khadilkar, M. R., Jiang, Y., Al-Dahhan, M., Dudukovic, M. P., Chou, S. K., Ahmed, G., et al. (1998). Investigations of a complex reaction network: I. Experiments in a high-pressure trickle bed reactor. *AIChE Journal, 44*(4), 912.

Khadilkar, M. R., Wu, Y. X., Al-Dahhan, M. H., & Dudukovic, M. P. (1996). Comparison of trickle bed and upflow reactor performance at high pressure: model prediction and experimental observations. *Chemical Engineering Science, 51*(10), 2139.

Kocis, G. R., & Ho, T. C. (1986). Effects of liquid evaporation on the performance of trickle-bed reactors. *Chemical Engineering Research and Design, 64*, 288−291.

Koros, R. M. (1986). Engineering aspects of trickle bed reactors. In H. De Lasa (Ed.), *Chemical reaction design and technology* (pp. 579−630). The Netherland: Martinus Nijoff; Cordrecht.

Korsten, H., & Hoffmann, U. (1996). Three phase reactor model for hydrotreating in pilot plant trickle-bed reactors. *AIChE Journal, 42*(5), 1350.

Kufner, R., & Hofmann, H. (1990). Implementation of radial porosity and velocity distribution in a reactor model for heterogeneous catalytic gas phase reactions. *Chemical Engineering Science, 45*(8), 2141−2146.

Lange, R., Gutsche, R., & Hanika, J. (1999). Forced periodic operation of a trickle-bed reactor. *Chemical Engineering Science, 54*.

Lee, H. H., & Smith, J. M. (1982). Trickle-bed reactors: criteria of negligible transport effects and of partial wetting. *Chemical Engineering Science, 37*(2), 223−227.

Lysova, A. A., Koptyug, I. V., Kulikov, A. V., Kirillov, V. A., Sagdeev, R. Z., & Parmon, V. N. (2007). Nuclear magnetic resonance imaging of an operating gas−liquid−solid catalytic fixed bed reactor. *Chemical Engineering Journal, 130*, 101−109.

Mears, D. E. (1974). The role of axial dispersion in trickle flow laboratory reactors. *Chemical Engineering Science, 26,* 1361–1366.

Merchan, A., Emig, G., Hofmann, H., & Chaudhari, R. V. (1986). NonEn Zur frage des katalysator-Wirkungsgrades bei folge Reactionen in mehrphasen-systemen. *Chemie Ingenieur Technik, 58,* 50.

Mills, P. L., & Chaudhari, R. V. (1997). Multiphase catalytic reactor engineering and design for pharmaceuticals and fine chemicals. *Catalysis Today, 37,* 367.

Mills, P. L., & Dudukovic, M. P. (1979). A dual-series solution for the effectiveness factor of partially wetted catalysts in trickle-bed reactors. *Industrial and Engineering Chemistry Fundamentals, 18,* 139.

Mills, P. L., & Dudukovic, M. P. (1980). Application of the method of weighted residuals to mixed boundary value problems – I. Dual series relations. *Chemical Engineering Science, 35*(7), 1557–1570.

Morita, S., & Smith, J. M. (1978). Mass transfer and contacting efficiency in a trickle-bed reactor. *Industrial and Engineering Chemistry Fundamentals, 17,* 113.

Morris Smith, C., & Satterfield, Charles N. (1986). Some effects of vapor–liquid flow ratio on performance of a trickle-bed reactor. *Chemical Engineering Science, 41*(4), 839–843.

Pintar, A., Bercic, G., & Levec, J. (1997). Catalytic liquid-phase oxidation of aqueous phenol solutions in a trickle bed reactor. *Chemical Engineering Science, 52,* 4143.

Rajashekharam, M. V., Jaganathan, R., & Chaudhari, R. V. (1998). A trickle-bed reactor model for hydrogenation of 2,4-dinitrotoluene: experimental verification. *Chemical Engineering Science, 53*(4), 787–805.

Ramachandran, P. A., & Chaudhari, R. V. (1983). *Three Phase Catalytic Reactors.* New York: Gordon and Breach.

Ramachandran, P. A., & Smith, J. M (1979a). Effectiveness factors in trickle-bed reactors. *AIChE Journal, 25*(3), 538.

Ramachandran, P. A., & Smith, J. M. (1979b). Mixing-cell method for design of trickle-bed reactors. *Chemical Engineering Journal, 17*(2), 91–99.

Ravindra, P. V., Rao, D. P., & Rao, M. S. (1997). A model for the oxidation of sulfur dioxide in a trickle bed reactor. *Industrial and Engineering Chemistry Research, 36,* 5125.

Roininen, J., Alopaeus, V., Toppinen, S., & Aitamma, J. (2009). Modeling and simulation of industrial trickle bed reactor for benzene hydrogenation: -Model validation against plant data. *Industrial and Engineering Chemistry Research, 48,* 1866.

Satterfield, C. N. (1975). Trickle-bed reactors. *AIChE Journal, 21,* 209–228.

Schnitzlein, K., & Hofmann, H. (1987). An alternative model for catalytic fixed bed reactors. *Chemical Engineering Science, 42,* 2569.

Sedricks, W., & Kenney, C. N. (1973). Partial wetting in trickle-bed reactors: the reduction of crotonaldehyde over a palladium catalyst. *Chemical Engineering Science, 28,* 559.

Shah, Y. T. (1979). *Gas–Liquid–Solid Reactor Design.* USA: McGraw-Hill.

Shah, Y. T., & Paraskos, J. A. (1975). Intraparticle diffusion effects in residue hydrodesulfurization. *Industrial and Engineering Chemistry Process Design and Development, 14*(4), 368–372.

Shah, Y. T., Stuart, E. B., & Sheth, K. D. (1976). Coke formation during thermal cracking of *n*-octane. *Industrial and Engineering Chemistry Process Design and Development, 15,* 518–524.

Sims, B. W., Gaskey, S. W., & Luss, D. (1994). Effect of flow regime and liquid velocity on conversion in a trickle bed reactor. *Industrial and Engineering Chemistry Research, 30,* 2530.

Stuber, F., Benaissa, M., & Delmas, H. (1995). Partial hydrogenation of 1,5,9-cyclododecatriene in three phase catalytic reactors. *Catalysis Today, 24,* 95.

Suwanprasop, S., Eftaxias, A., Stuber, F., Polaert, I., Julcour, C., & Delmas, H. (2005). Scaleup and modeling of fixed bed reactors for catalytic phenol oxidation over adsorptive active carbon. *Industrial and Engineering Chemistry Research, 44*, 9513.

Tan, C. S., & Smith, J. M. (1980). Catalyst particle effectiveness with unsymmetrical boundary conditions. *Chemical Engineering Science, 35*, 1601.

Tukac, V., & Hanika, J. (1998). Catalytic wet oxidation of substituted phenols in the trickle bed reactor. *Chemical Engineering Technology, 21*, 262.

Valerius, G., Zhu, X., Hofmann, H., Arntz, D., & Haas, T. (1996). Modeling of a trickle-bed reactor II. The hydrogenation of 3-hydroxypropanal to 1,3-propanediol. *Chemical Engineering and Processing, 35*, 11.

Vergel, C., Euzen, J. P., Trambouze, P., & Wauquier, J. P. (1995). Two-phase flow catalytic reactors, influence of hydrodynamics on selectivity. *Chemical Engineering Science, 50*, 3303.

Wehner, J. F., & Wilhelm, R. H. (1956). Boundary conditions of flow reactor. *Chemical Engineering Science, 6*, 89—93.

Westerterp, K. R., & Wammes, W. J. A. (2000), Three-phase trickle-bed reactors. In: Ullmann's encyclopedia of industrial chemistry. John Wiley and Sons Inc.

Wu, Q., Hu, X., Yue, P., Feng, J., Chen, X., Zhang, H., & Qiao, S. (2009). Modeling of a pilot scale trickle bed reactor for the catalytic oxidation of phenol. *Separation and Purification Technology, 67*, 158.

Yamada, Hiroshi, & Goto, S. (1997). Periodic operation of trickle bed reactor for hydrogenolysis in gas—liquid—liquid—solid four phases. *Journal of Chemical Engineering of Japan, 30*(3), 478—483.

Chapter 4

Flow Modeling of Trickle Beds

Everything should be made as simple as possible, but not simpler.

Albert Einstein

INTRODUCTION

Conventional approaches for the analysis of trickle bed reactors are discussed in previous two chapters. Gross hydrodynamic characteristics of trickle bed reactors and different correlations for estimating key design parameters are summarized in Chapter 2. The interaction of downward flow of gas and liquid phases with closely packed solid particles leads to different flow regimes in trickle bed reactors. The classification of these flow regimes, their key characteristics, and influence on other important design parameters are also discussed in Chapter 2. The flow regime information is used for estimation of design parameters in the classical reaction engineering analysis for simulation of overall reactor performance as discussed in Chapter 3. The basic methodology discussed in these two previous chapters may not, however, be sufficient to establish quantitative relationship between reactor hardware/operating protocols and reactor performance, particularly at different scales of operation. The primary reason for this is these types of models do not adequately account for interaction of the hydrodynamics of the multiphase systems involved with various transport and kinetic processes occurring on different scales (from molecular scale to tens of meters). Such interactions over a wide range of scales cause severe difficulties in developing and solving predictive models to simulate reactor performance. It is often difficult to develop a single model to describe the complete reactor performance for a complex system such as a trickle bed reactor. Therefore, it is necessary to develop multilayer or multiscale models, which comprise of multiple models, each simulating processes occurring at different scales with different objectives and level of complexity. These models communicate the information with each other to provide adequately accurate estimation of the overall reactor performance. Such an approach for reactor engineering of trickle bed reactors is described here.

Trickle Bed Reactors. DOI: 10.1016/B978-0-444-52738-7.10001-4

Development of a trickle bed reactor design involves several steps such as:

- design or selection of a catalyst (composition, support, particle size/shape, surface area, porosity, etc.)
- development of rate equations and evaluation of kinetic parameters (understanding key chemical reactions, estimating intrinsic rate constants and activation energies, influence of catalyst conditioning, deactivation kinetics, etc.)
- evaluation of interphase and intraparticle (diffusion in porous catalysts; single component and multicomponent) mass transfer parameters under conditions of different flow regimes
- evaluation of key hydrodynamic characteristics (wettability, packing density, pressure drop, gas and liquid holdup, etc.)
- evaluation of scale-up/scale-down-related aspects (maldistribution, channeling, heat transfer, hotspots, residence time distribution (RTD), etc.)
- fine tuning scale-up and design of large-scale reactor (distributors, inlet–outlet nozzles, any other internals, location of temperature sensors, etc.)

The first four steps can be executed using the methods discussed in previous two chapters. These steps will lead to a preliminary design of a trickle bed reactor. Converting this preliminary design to a level that will be applicable at different scales of operation requires execution of the last two steps. Conventionally, the scale-up issues (the last two steps) are addressed based on the experience and engineering judgment of design engineers. In this chapter, an attempt will be made to develop a comprehensive multiscale analysis methodology based on a sound scientific approach. A brief introduction to the elementary processes occurring on different scales in a trickle bed reactor is given in Chapter 1 (see Figs. 5 and 6 and associated discussion). The processes occurring within the catalyst pores and on active catalyst sites can be quantitatively analyzed using the methods discussed in Chapter 3. In this chapter, methods for analysis of processes occurring on scales of individual catalyst particles, group of particles (meso-scale processes), and catalyst bed (macro-scale processes) in a trickle bed reactor are discussed. When integrated with the issues discussed in Chapters 2 and 3, these methods will provide means to implement steps listed above, involved in the development of trickle bed reactor design. Some other practical aspects of reactor engineering of trickle beds are discussed in Chapter 5.

In this chapter, key issues like characterization of individual catalyst particle, packing density, and packed bed of such particles are discussed. Different possible arrangements of catalyst particles in the bed and their influence on other parameters are discussed in the next section. Mathematical ways of constructing and quantitatively characterizing packed beds are briefly discussed. Single-phase flow through group of particles is then discussed in the section *Single-Phase Flow Through Packed Bed*. It provides basic information on drag exerted by a fluid on particles, flow structures in packed bed, local heat and mass transfer parameters, etc. Gas—liquid flow through packed bed is discussed under the section *Gas—Liquid Flow Through Packed Bed*. Computational models to simulate gas—liquid flow through large packed beds of catalyst particles are discussed in this section. Meso-scale models which may provide better insight into wetting and interaction of liquid phase with solid particles are also briefly discussed. Applications of these flow models for simulating RTD and chemical reactions in trickle bed reactors are then discussed. An attempt is made to provide a general framework and guidelines for using different computational models for constructing as complete picture of the large-scale trickle bed reactor as possible.

CHARACTERIZATION OF PACKED BEDS

The main objective in designing trickle bed reactors is to improve effective contact of gas and liquid phase reactants with the active catalyst sites; provide adequate residence time and allow dissipation of liberated heat of reactions without causing any hotspots and catalyst deactivation. Usually, the following factors are considered while designing a packed bed for trickle flow operation:

Pressure drop: Operating cost of trickle bed reactors is directly related to pressure drop across the bed and is crucially dependent on the structure of the bed and operating flow rates. It is possible to manipulate the bed structure (particle characteristics and packing characteristics) to reduce pressure drop without jeopardizing the overall performance of the reactor. Maintaining a stable pressure drop is also essential for smooth operation of the reactor for prolonged period of time.

Specific surface area: Effective contacting of reactants with active catalyst site and heat/mass transfer efficiency are determined by available specific surface area as one of the factors.

Residence time distribution/mixing: It is essential to understand the relationship between bed structure and resulting RTD in order to achieve the desired RTD by manipulating the bed structure. Catalyst particle size, shape, and surface characteristics (to manipulate wetting) influence RTD. Maldistribution and channeling of the liquid or gas flow, which adversely influence the overall reactor performance, depend on bed structure and method of packing. Liquid and gas phase axial and radial mixing is also influenced by local bed structure and associated meso-scale flow processes. The existence of dead zones (stagnant pockets) in the reactor reduces effective utilization of packed catalyst. The bed structure needs to be designed to minimize if not eliminate such dead zones.

Liquid holdup: Higher liquid holdup is usually beneficial for efficient mass transfer and higher reaction rates and therefore the bed structure needs to be manipulated to achieve higher liquid holdup.

Heat and mass transfer: Particle shape, size, and local arrangements influence local flow structure and therefore local heat and mass transfer rates.

These local transport rates become important when the reaction is mass transfer limiting or temperature sensitive.

It is obvious that these different aspects may impose conflicting demands on bed structure. For example, a structure that provides higher heat and mass transfer rates usually leads to higher pressure drop. It is therefore essential to consider these different aspects and evolve a compromise solution appropriate for a specific case under consideration.

The physical nature of the catalyst surface on which chemical transformations occur is quite crucial. Active catalytic material consisting of metals/metal oxides or their complexes is deposited on the surface of a carrier material (alumina, silica, or specially designed zeolites or metallic surface). In most cases, catalyst particles are porous and the active material is impregnated on the surface of internal pores. Chemical reactions occur on catalyst sites available on these micro-pores within the catalyst particles. In some cases, to avoid diffusional resistances, catalyst is impregnated on outer surface (or shell) instead of impregnating inside the entire particles. This is called eggshell-type catalyst particles. If reaction is highly exothermic, non-porous catalyst particles are used where reaction occurs only on the external surface of the particles. Sometimes, catalyst is coated on a structured support instead of particles. Apart from the distribution of chemically active sites, physical nature of the catalyst comprising size, shape of particle, density, surface area per unit volume, surface characteristics (roughness wetting), heat capacity, thermal conductivity, attrition resistance, and strength to compression load are also important. The effective catalyst life is dependent on possibility of leaching, thermal deactivation, or sintering, etc. Historically, spherical and cylindrical-shaped porous catalyst particles are widely used. In recent years, several new shapes of catalyst particles have been proposed for industrial applications (see Fig. 1 of Chapter 1 and discussion in the section *Particle Characteristics* of Chapter 5).

After selection of a particular type of catalyst, next important task is to design the nature of packed bed and packing structure. Most widely used practice is to pack the catalyst particles randomly inside the bed as shown in Fig. 1a. However, several other ways of realizing semi-structured and structured packing of catalyst particles have also been proposed (Irandoust & Anderson, 1988; Mewes, Loser, & Millies, 1999; Van Hasselt, Lindenbergh, Calis, Sie, & Van Den Bleek, 1997). A schematic of structured packing arrangement is shown in Fig. 1b. A brief discussion on characteristics of random and structured beds is provided in the following sections.

Randomly Packed Bed

Randomly packed beds are used in majority of industrially operated trickle bed reactors because of their simplicity in construction and loading procedure. Randomly packed catalyst particles are generally spherical, cylindrical,

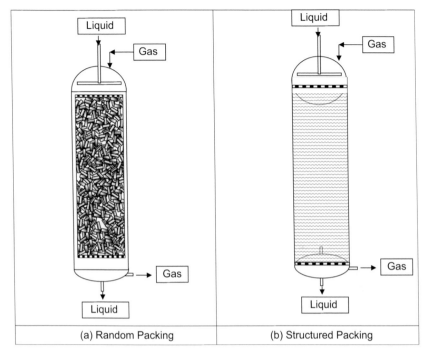

(a) Random Packing	(b) Structured Packing

FIGURE 1   A schematic of random and structured packing. (a) Random packing; (b) structured packing.

extrudates, trilobes, or quadrilobes. Selection of particle size and shape is often based on the desired pressure drop in the system as well as other transport characteristics like external and intraparticle mass and heat transfer. Attrition and durability (stiffness) of particles are also important parameters for selecting the catalysts. Catalyst attrition may occur during loading of catalysts which may lead to a higher pressure drop and loss of catalyst. There are several methods for realizing such a random packing in practice as described by Al-Dahhan, Wu, and Dudukovic (1995). When particles are packed randomly in a cylinder, the characteristic of packing depends significantly on the ratio of particle size to the bed diameter and on the shape of particles.

Numerous studies on porosity distribution in randomly packed beds are available (Donohue & Wensrich, 2008; Mantle, Sederman, & Gladden, 2001; Spedding & Spencer, 1995; Stephenson & Stewart, 1986). These experimental and computational studies have shown that the bed porosity is higher near the vicinity of the reactor wall and it fluctuates significantly in the near wall region (of width of about 4–5 particle diameters). Mueller (1991) has proposed a correlation for radial variation of axially averaged porosity as a function of column diameter, particle diameter, and average porosity. This correlation (Eq. (1)) represents the available experimental data with a reasonable accuracy:

$$\varepsilon(r) = \varepsilon_B + \left(1 - \varepsilon_B\right)J_0\left(ar^*\right)e^{-br}.$$

where,

$$a = 8.243 - \frac{12.98}{\left(D/d_p - 3.156\right)} \quad \text{for } 2.61 \leq D/d_p \leq 13.0$$

$$a = 7.383 - \frac{2.932}{\left(D/d_p - 9.864\right)} \quad \text{for } 13.0 \leq D/d_p$$

$$b = 0.304 - \frac{0.724}{D/d_P}$$

$$r^* = r/D \quad \text{and} \quad J_0 \text{ is zero-th order Bessel function} \tag{1}$$

The predicted radial variation of bed porosity for 11.4 cm column diameter for two particle sizes is shown in Fig. 2 as an illustration.

Jiang, Khadilkar, Al-Dahhan, and Dudukovic (2001) have shown that the porosity variation in axial direction (at any radial location) is close to the Gaussian distribution. The value of standard deviation of such a distribution decreases with increase in the ratio of column diameter to particle diameter and eventually approaches to zero for very small (compared to column diameter) particles. The porosity distribution within the randomly packed bed may thus be represented by imposing random fluctuations (following the Gaussian distribution) over the axially averaged porosity estimated by Mueller's correlation (see,

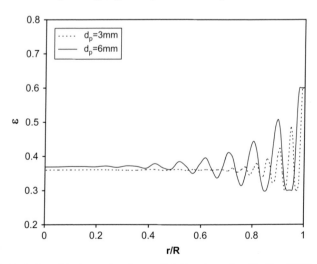

FIGURE 2 Variation of bed porosity along radial direction using Mueller (1991) correlation for column diameter of 11.4 cm (Gunjalet al., 2005). System: $D = 0.114$ m, 3 mm or 6 mm glass beads.

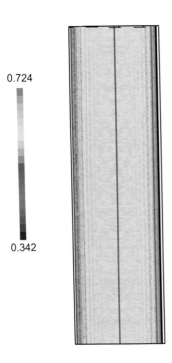

0.724

0.342

FIGURE 3 Contours of radial and axial variation of solid holdup implemented in computational model (Gunjal et al., 2005). System: $D = 0.194$ m, 6 mm glass beads.

for example, Gunjal, Kashid, Ranade, & Chaudhari, 2005). A sample of random solid phase distribution as a result of implementation of porosity distribution generated by the above method is shown in Fig. 3. Such randomly distributed bed porosity may give more realistic results than assuming mean porosity all over the bed. It must be noted that the porosity distribution observed in a packed bed will be obviously dependent on the scale of scrutiny. It has been experimentally shown that on the scale of a cluster of particles, porosity has the Gaussian distribution (Jiang, Khadilkar, Al-Dahhan, & Dudukovic, 2000) while at a much smaller scale porosity has a bi-modal distribution (Jiang et al., 2001). This relationship between porosity distribution and scale of scrutiny (scale of interest) should be kept in mind while generating randomly distributed porosity within a bed.

Most of the information available in open literature is, however, restricted to spherical particles. Comparison of a radial variation of porosity with trilobe and spherical particles is shown in Fig. 4. It can be seen that fluctuations in porosity with trilobes are less than those observed with the spherical particles. In the absence of adequate information on porosity distribution for catalyst particles of different shapes, discrete element methodology may be used to quantify characteristics of the packed beds (Donohue & Wensrich, 2008). These computational models simulate random packing of the bed and provide a useful framework to quantitatively understand the influence of particle size and shape on characteristics of the randomly packed beds. A sample of such simulated bed packing is shown in Fig. 5 (Donohue & Wensrich, 2008).

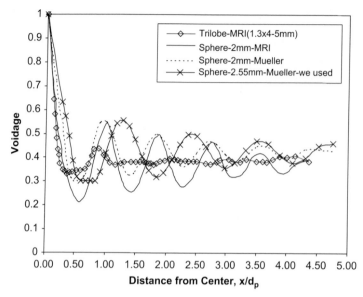

FIGURE 4   Variation of porosity for different catalyst sizes and shapes (Gunjal & Ranade, 2007). System: $D = 0.0224$ m, $\varepsilon = 0.5$.

Structured Bed

In order to avoid likely build up of higher pressure drop associated with the random packing, attempts have been made to use semi-structured or fully structured packed beds for trickle bed reactors. The ideas about the structured packing mainly originated from their successful application in distillation columns. High surface area per unit volume, lower pressure drop, and relatively straightforward scale-up are some of the key advantages of the structured packing. Therefore, structured packing may be preferred over the random unstructured packing for processes requiring higher mass transfer rates and lower pressure drop. However, structured packing is expensive compared to the conventional unstructured packing and often needs additional support mechanisms for installation inside the reactors. Several types of structured packing as listed below may be used in the trickle bed reactors:

- Gauze packing
- Corrugated sheet packing
- Mesh-type packing
- Monoliths/honeycomb (Irandoust & Anderson, 1988)
- Three layer packing (Van Hasselt et al., 1997)

Structured packing can be incorporated in reactors by several ways; for example, arranged in continuous stacked form with or without offset, structured packing incorporated in a pipe, and arranged on the sieve plates. Structured

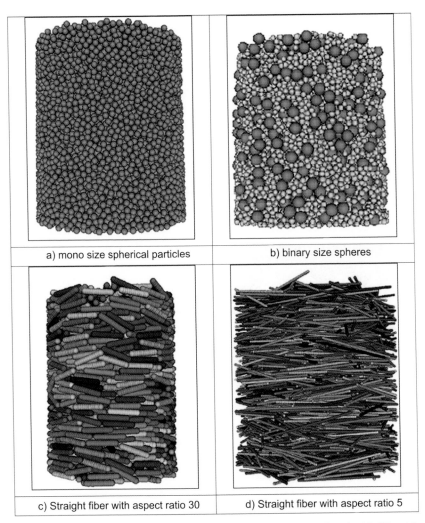

FIGURE 5 Simulated packed beds with particles of different shapes (Donohue & Wensrich, 2008). (a) Mono size spherical particles; (b) binary size spheres; (c) straight fiber with aspect ratio 30; (d) straight fiber with aspect ratio 5.

packing is characterized in terms of characteristic scale of repeat unit, surface area per unit volume, and void fraction. Some examples of the structured packing are shown in Fig. 6.

It is important to accurately quantify characteristics of the packed bed (characteristic length scales and porosity distribution) to facilitate better understanding of the flow behavior of gas and liquid phases. The available information and models for obtaining insight into flow through trickle beds are discussed in the following.

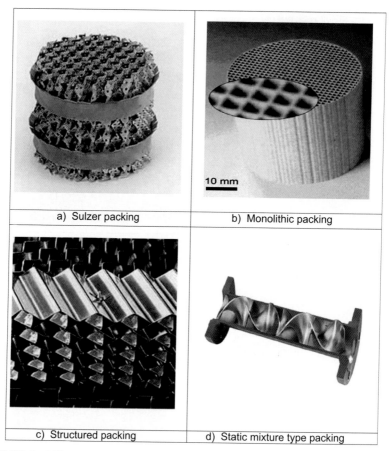

| a) Sulzer packing | b) Monolithic packing |
| c) Structured packing | d) Static mixture type packing |

FIGURE 6 Different types of structured packing used in trickle bed reactors (http://www. sulzerchemtech.com/). (a) Sulzer packing; (b) monolithic packing; (c) structured packing; (d) static mixture-type packing.

SINGLE-PHASE FLOW THROUGH PACKED BED

In trickle bed reactors, gas and liquid phases flow cocurrently downward through the packed bed of catalyst particles. Multiphase flow through complex interstitial geometry formed by bed particles controls mixing and other transport processes occurring in trickle bed reactors. It is important to gain insight into the local flow behavior and transport processes occurring in the packed bed to be able to tailor, monitor, and optimize its performance. Considering the complexities of the multiphase flow through packed bed, it is necessary to first develop a clear understanding of the single-phase flow inside the bed before proceeding to the analysis of multiphase flow behavior. Some of the key aspects of single-phase flow through packed bed and the various approaches of computational modeling have been discussed in this part to create a platform for further discussion on gas—liquid flow through packed beds.

Besides its importance for understanding the multiphase flow behavior, single-phase flow through the packed bed is also encountered in a variety of applications ranging from flow through capillary column of a gas chromatograph to flow through pebbles in the riverbed. Single-phase flow occurs via void region present in the bed consisting of interconnected pores and tortuous path. In trickle bed reactors, the catalyst bed is usually filled with porous catalyst particles. The transport inside such porous particles is mainly controlled by diffusional processes and therefore is not significantly influenced by flow through voids formed due to packing of particles. However, the mass and heat transfer external to the catalyst particles and axial as well as radial mixing can be significantly affected by the changes in the flow behavior of the gas and liquid phases through the bed. Flow through voids of packed bed has been extensively studied using experimental as well as computational models. Prevailing regimes of single-phase flow through packed beds depending on the gas and liquid velocities are briefly discussed in Chapter 2. Key aspects of different regimes of single-phase flow through packed beds are highlighted in the following.

Sangani and Acrivos (1982) and SØrensen and Stewart (1974) have analyzed the flow through unit cells with particles arranged in simple cubical (SC) and face centered cubic (FCC) patterns several years ago. They have reported predicted drag force exerted on particles at different values of porosity. Their studies, however, were limited to the Stokes flow regime ($Re \rightarrow 0$). The Stokes flow regime exists at very low particle the Reynolds number ($\sim<0.1$) where average drag force on particles is independent of Reynolds number. Considering the usual operating ranges of packed beds, it is, however, essential to understand the flow characteristics in the inertial flow regime (particle Reynolds number > 10). Only recently some attempts of analyzing inertial flow in packed beds have been made.

Durst, Haas, and Interthal (1987) simulated laminar flow through unit cells and compared simulated pressure drop with experimental data. Maier, Kroll, Kutovsky, Davis, and Bernard (1998) have carried out lattice Boltzmann simulations of single-phase flow through FCC and random packing arrangement of particles. They presented some comparisons of simulated velocity distribution with their experimental results. Their study was restricted to low Reynolds numbers ($Re = 0.5-29$). Tobis (2000) has used unit cell approach for simulating turbulent flow in a packed bed. Hill, Koch, and Ladd (2001) have carried out the lattice Boltzmann simulations of flow through FCC, SC, and random arrangement of particles. Detailed analysis of drag force variation with solid volume fraction and packing arrangement was discussed. Freund et al. (2003) have carried out the lattice Boltzmann simulations of flow in a packed bed reactor of a low aspect ratio (~5). They studied relative contributions of skin and form drag in overall pressure drop. However, in these studies, detailed analysis of flow structure in a packed bed and its influence on rates of other transport processes was not presented. Recently, Magnico (2003) has carried out simulations for a unit cell and for small tube-to-sphere diameter ratio over the range of Reynolds numbers (from 7 to 200). He demonstrated the influence of flow structures on mass transfer with the help of the Lagrangian particle tracking.

Extensive experimental data of flow through packed beds are now available. Mantle et al. (2001), Sederman and Gladden (2001), and Sederman, Johns, Bramley, Alexander, and Gladden (1997) have experimentally characterized flow in packed beds and have reported measured distributions of axial and transverse velocities in the interstitial space. The axial velocity distribution showed a sharp peak and was asymmetrical. The transverse velocity distribution showed exponential decay in both positive and negative directions. Experimental data reported by Maier et al. (1998) also showed similar trends of velocity distributions. Suekane, Yokouchi, and Hirai (2003) have carried out detailed measurements of flow through voids of simple packed bed using magnetic resonance imaging (MRI) and have provided detailed quantitative data on velocity distributions. They also reported details of inertial flow structures for different Reynolds numbers. It is more effective to use such comprehensive data sets to validate computational models and use the validated models for further investigations. Several attempts have been made recently to develop computational models for simulating the flow behavior (flow structures, velocity field) in packed beds (see, for example, Calis, Nijenhuis, Paikert, Dautzenberg, & van den Bleek, 2001; Dixon & Nijemeisland, 2001; Freund et al., 2003; Logtenberg, Nijemeisland, & Dixon, 1999; Tobis, 2000; Zeiser et al., 2002).

Apart from such detailed studies of meso-scale flow, computational models have also been used to quantify macroscopic flow characteristics such as maldistribution or channeling in packed beds (mixing and RTDs). In such cases, the entire bed is modeled as a porous medium. The local porosity values are specified by averaging over a control volume, significantly larger than the particle size. Before describing a sample of results and key aspects of single-phase flow through packed beds, modeling approaches used for such studies are briefly reviewed in the following.

Modeling Approaches

Flow through a packed bed can be modeled using various approaches depending on the objectives and intended use. These approaches are schematically shown in Fig. 7.

In all of these approaches mass and momentum balance equations are solved to simulate detailed flow characteristics around particles/bed of particles. In the first approach (A), entire packed bed consisting of a number of particles (either arranged in a regular fashion or in a random fashion) is

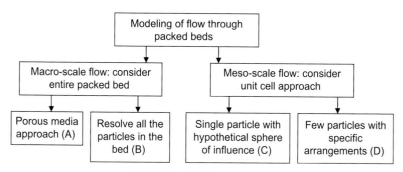

FIGURE 7   Modeling approaches.

considered. Porous media approach (A), however, does not provide information about the local flow and transport occurring within the bed. It provides the macroscopic information. Other three approaches (B–D) can provide local information (on particle scale) and are discussed here. Calis et al. (2001), Logtenberg et al. (1999), and Nijemeisland and Dixon (2001) among others have used the approach (B). Computational constraints often limit the size of the bed and the number of particles considered in such simulations. Therefore, often the "unit cell approaches" (C, D) are used.

As shown in Fig. 7, there are two sub-types in the unit cell approach. In the first type (C), each particle is assumed to have a hypothetical sphere of influence around it (see Dhole, Chhabra, & Eswaran, 2004 and references cited therein). Flow is solved around a particle placed in a hypothetical sphere of influence (size of which depends on porosity of the bed). This approach, however, ignores differences caused by different particle arrangements and therefore was not considered here. In the second type of unit cell approach (D), a unit "periodic" cell composing of few particles is considered. The packed bed is represented by periodically repeating the unit cell in all the three directions. This approach is being used traditionally to analyze transport processes in packed beds (see, for example, Martin, McCabe, & Monrad, 1951; SØrensen & Stewart, 1974). Different packing arrangements of particles like SC, rhombohedral, and FCC or body centered cubic (BCC) can be considered for representing the packed bed. This approach is illustrated here.

The approach of unit cells, where packed bed of spheres is represented by geometrically periodic unit cells with different packing arrangements, is advantageous to understand flow structures in large packed beds. It is, however, essential to understand the possible implications of approximating a packed bed by periodic unit cells. It is well known that symmetry of a flow over a single sphere breaks when particle Reynolds number increases beyond 105 and unsteady flow occurs (Natarajan & Acrivos, 1993). The unit cell approach is not valid for cases where periodic symmetry of flow is absent despite the symmetric and periodic geometry. Fortunately, when particles are packed closely together in a regular fashion, the onset of symmetry breaking unsteady flow is delayed considerably (Hill et al., 2001). The largest length scale characterizing the interstitial region of the regular arrays is smaller than particle diameter. Therefore, the Reynolds number characterizing the stability of the flow in the interstitial region can be up to approximately 2.5 times larger than the critical particle Reynolds number. Secondly, at larger solid volume fractions, the fluid is increasingly confined and hence stabilized by neighboring spheres. For a specific particle Reynolds number, viscous dissipation will be higher at higher solid volume fraction, and therefore more effective in damping velocity fluctuations. Considering this, the unit cell approach can be used to understand the influence of particle Reynolds number and packing arrangement on inertial flow structures in packed beds.

Model Equations and Boundary Conditions

For approaches (B), (C), and (D), classical single-phase flow equations with appropriate selection of solution domain and boundary conditions can be used. Laminar flow of an incompressible fluid through a packed bed of spheres can be simulated by solving the Navier–Stokes equations. The study of Seguin, Montillet, and Comiti (1998) and Seguin, Montillet, Comiti, and Huet (1998) indicated that the flow in packed bed exhibits a transition regime over a large range of particle Reynolds number and the turbulent flow regime may exist beyond $Re_p = 900$. Therefore, when particle Reynolds number exceeds 1000, the Reynolds-averaged Navier–Stokes equations (RANS) along with suitable turbulence model need to be used (see Ranade, 2002 for more discussion on modeling of turbulent flows). Reynolds-averaged Navier–Stokes equations for mass and momentum balance for incompressible Newtonian fluid are given by,

$$\frac{\partial u_i}{\partial x_i} = 0 \tag{2}$$

$$\frac{\partial u_i}{\partial t} + \frac{\partial (u_i u_j)}{\partial x_j} = -\frac{1}{\rho}\frac{\partial P}{\partial x_i} + \frac{\partial}{\partial x_j}\left[(\nu + \nu_t)\left(\frac{\partial u_i}{\partial x_j} + \frac{\partial u_j}{\partial x_i}\right)\right] \tag{3}$$

where, u_i is the mean velocity in the i direction, P is the pressure and ν_t is the turbulent kinematic viscosity. The turbulent kinematic viscosity, ν_t, needs to be estimated using an appropriate turbulence model. The two-equation standard $k-\varepsilon$ model of turbulence is one of the most widely used turbulence models and can be used as a starting point to simulate turbulent flow in packed beds. The governing equations of this model can be written as:

$$\nu_t = C_\mu \frac{k^2}{\varepsilon} \tag{4}$$

$$\frac{Dk}{Dt} = \frac{\partial}{\partial x_j}\left[\left(\nu + \frac{\nu_t}{\sigma_k}\right)\cdot\frac{\partial k}{\partial x_j}\right] + G - \varepsilon \quad G = \nu_t\frac{\partial u_i}{\partial x_j}\left(\frac{\partial u_i}{\partial x_j} + \frac{\partial u_j}{\partial x_i}\right) \tag{5}$$

$$\frac{D\varepsilon}{Dt} = \frac{\partial}{\partial x_j}\left[\left(\nu + \frac{\nu_t}{\sigma_\varepsilon}\right)\cdot\frac{\partial\varepsilon}{\partial x_j}\right] + \frac{\varepsilon}{k}(C_{1\varepsilon}G - C_{2\varepsilon}\varepsilon) \tag{6}$$

The standard values of the parameters appearing in Eqs. (5) and (6) were used ($C_{1\varepsilon} = 1.44$, $C_{2\varepsilon} = 1.92$, $C_\mu = 0.09$, $\sigma_\kappa = 1.0$, and $\sigma_\varepsilon = 1.3$: from Launder & Spalding, 1974).

The energy conservation equation (without any source or sink) may be written as:

$$\frac{\partial(\rho h)}{\partial t} + \nabla \cdot (\rho u_i h) = (k + k_t)\nabla^2 T \tag{7}$$

where h is enthalpy, k is thermal conductivity, and k_t is turbulent thermal conductivity. First term in the right-hand side is of the effective conductive heat flux (molecular and turbulent). The turbulent thermal conductivity is given by,

$$k_t = \frac{C_p\mu_t}{Pr_t} \tag{8}$$

In order to model the unit cell as a representative piece of a packed bed, periodic boundary conditions need to be implemented at all the faces of the unit cell through which flow occurs. In such a translational periodic boundary condition, all the variables except pressure (and enthalpy or temperature) at periodic planes are set to be equal. For a desired particle Reynolds number, superficial velocity and mass flow rates were calculated based on the considered geometry. However, pressure is not periodic; instead the pressure drop is periodic. Because the value of pressure gradient is not known a priori, it must be iterated until the specified mass flow rate is achieved in the computational model. No slip boundary condition needs to be implemented on all the impermeable walls. For turbulent flow simulations, it may not be possible to resolve the steep gradients near walls without excessively increasing demands on computing resources. In such cases, a concept of "wall functions" is used (standard wall function of Launder & Spalding, 1974). Several extensions of the basic wall functions have been developed (see, for example, enhanced wall function by Jongen, 1992; Kader, 1993; Wolfstein, 1969; non-equilibrium wall function by Kim & Choudhury, 1995; generalized wall functions by Popovac & Hanjalic, 1997). For carrying out heat transfer simulations; if the interest is to simulate heat transfer coefficient, it may be necessary to resolve the steep gradients near the walls adequately. The heat transfer coefficient from solid wall to the fluid can then be calculated as;

$$q = hA(T_w - T_b) = -k_f\left(\frac{\partial T}{\partial n}\right)_{wall} \tag{9}$$

where, n is the normal coordinate normal to the wall and k_f is the thermal conductivity of the fluid phase. The subscripts w and b denote wall and bulk, respectively.

Approaches (B), (C), and (D) are more suitable as learning models or as supportive tools for validation of data collected experimentally. This may also reduce use of empirical parameters while modeling actual reactor. For example, local particle to fluid heat transfer coefficient can be calculated with high accuracy using the above approach for known flow rates and particle type. The results obtained based on the assumption of periodic array need to be interpreted and connected to the flow in large bed of particles. This relationship is not straightforward and requires characterization of different possible packing arrangements in the large bed. In the absence of such characterization, the approach (A) is more suitable in practice for simulating flow through large packed beds. In this approach, instead of actually simulating the packing geometry inside the column and solving governing equation of single-phase flow in the void spaces inside the reactor, bed of randomly packed particles is considered as a continuous porous medium with predefined porosity distribution. This assumes that definition of a volume fraction of solids (over a suitable control volume) is meaningful. Continuity and momentum balance equations are derived using appropriate averaging. Additional source terms representing resistance offered by the porous medium (randomly packed bed) are included in the momentum equations. More details of this approach and governing equations may be found elsewhere (Ranade, 2002). Unlike the other three modeling approaches, the approach (A) can be applied to industrial-scale packed beds. However, in the formulation of the averaged governing equations, some semi-empirical interaction and closure terms are involved. Hence, supportive experiments and models as well as learning models (direct simulation or periodic assumption approach) are required. Some of the results obtained with the approaches discussed here are described in the following section.

Flow Through an Array of Particles

Different types of flow patterns were observed during single-phase flow through periodic array of particles depending upon the flow rates, properties of fluids, and loading of solid. Mantle et al. (2001), Sederman and Gladden (2001), and Sederman et al. (1997) have experimentally characterized flow in packed beds and have reported measured distributions of axial and transverse velocities in interstitial space. They also reported details of inertial flow structures for different Reynolds numbers. Gunjal, Chaudhari, and Ranade (2005a) have validated their computational model using the data of Suekane et al. (2003). Some of these results are reproduced here for providing glimpses of key characteristics of flow through array of particles.

Simple Cubic Arrangement of Spheres

It is important to ensure that predicted results are not influenced by numerical parameters like discretization schemes and grid size when computational models described in the previous section are used for flow simulations. Higher order discretization schemes are recommended. It is also useful to carry out numerical experiments to quantify influence of grid size, distribution, and discretization schemes on predicted results. Comparison of results predicted by Gunjal et al. (2005a) with the experimental data of Suekane et al. (2003) is shown in Fig. 8.

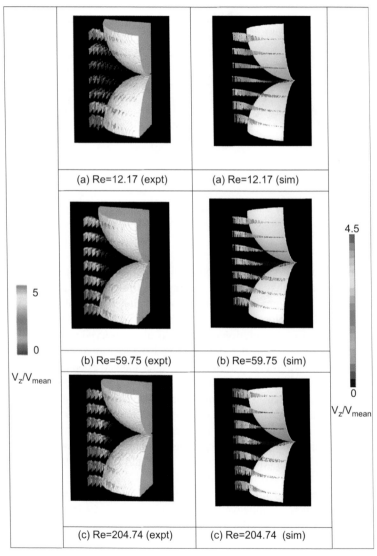

FIGURE 8 Comparison of simulated results of z-velocity (axial velocity) with experimental data at different particle Reynolds numbers (Gunjal et al., 2005a). (a) $Re = 12.17$; (b) $Re = 59.75$; (c) $Re = 204.74$.

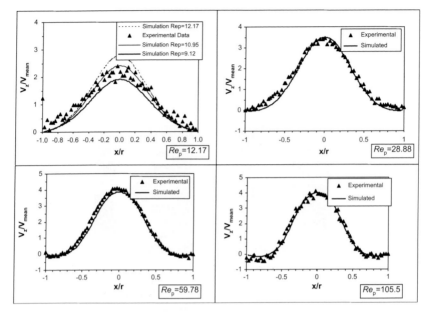

FIGURE 9 Comparison of simulated z-velocity (axial velocity) distribution with experimental data at various Reynolds numbers. Velocity measured at z midplane along x-axis at $y = 0.0015$ (Gunjal et al., 2005a).

It can be seen that variation of axial velocity was captured reasonably well in the simulated results. At highest Reynolds number ($Re_p = 204.74$), where inertial forces are dominant, jet-like flow behavior was observed in the experimental flow fields (see Fig. 8c). Similar, dominant velocity stream through the center of the solution domain was also observed in the simulation. Quantitative comparison of the simulated and the measured z-components of the velocity is shown in Fig. 9. Simulated results showed good agreement with the experimental data except at the lowest value of Reynolds number (12.17). It is interesting to note that experimental results also show highest scatter at this particle Reynolds number. Possible difficulties in maintaining a steady flow at a very low flow rate may be one of the reasons for such scatter. It can be seen (from Fig. 9) that the reported experimental data for the lowest Reynolds number (12.17) lie in between the results predicted for $Re_p = 9.12$ (25% less than 12.17) and 10.95 (10% less than 12.17).

Simulated results of flow field were also found to capture experimentally observed flow structures quite well (see Fig. 10 for $Re_p = 59.78$ and Fig. 11 for $Re_p = 204.74$). It is noteworthy that at higher Reynolds number (204.74), the observed and simulated flows at plane C are qualitatively different than those observed at lower Reynolds number (59.78). Further simulations at higher particle Reynolds number show that the normalized velocity profiles are almost independent of Reynolds number unlike in the laminar flow regime (where

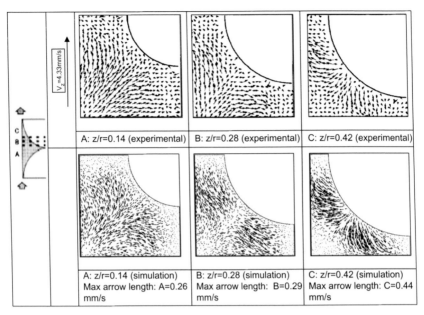

FIGURE 10 Comparison of the simulated flow field at three horizontal planes with experimental data at $Re_p = 59.78$ (Gunjal et al., 2005a).

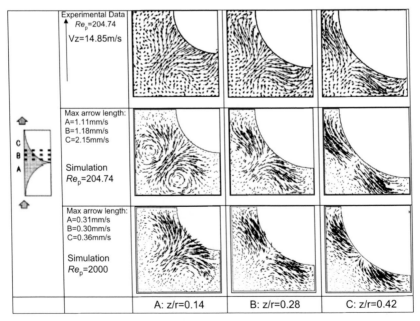

FIGURE 11 Comparison of the simulated flow field with experimental data at three horizontal planes ($Re_p = 204.74$ and 2000).

maximum in profiles of normalized axial velocities increases with increasing Reynolds number). The maximum value of normalized velocity in turbulent regime is closer to that obtained with the lowest Reynolds number considered in the laminar regime ($Re_p = 12.17$). Velocity profiles for the turbulence cases are much flatter than those obtained for the laminar regime. The region of negative velocities near the wall was found to be larger in turbulent flow regime compared to the laminar regime. Inertial flow structures were similar to those observed at particle Reynolds number of 204.74. Knowledge of such detailed flow structure within void spaces will have significant implications for the estimation of heat and mass transfer parameters. For example, simulated results of Gunjal et al. (2005a) indicate substantial difference in the predicted Nusselt numbers for the FCC arrangement and SC arrangement of spheres. Influence of particle arrangement on resulting flow structures and velocity distribution is briefly discussed in the following section.

Influence of Packing Arrangement of Spheres

In order to understand the influence of packing arrangement of particles in unit cell, results of Gunjal et al. (2005a) are discussed here. These simulations were carried out for different particle arrangements: namely one-dimensional rhombohedral ($\varepsilon = 0.4547$), three-dimensional rhombohedral ($\varepsilon = 0.2595$), and FCC ($\varepsilon = 0.302$) arrangements. Simulations were carried out at different particle Reynolds numbers in laminar flow regime ($Re_p = 12.17-204.74$) and in turbulent flow regime ($Re_p = 1000, 2000$).

Simulated results for one-dimensional rhombohedral arrangement at different particle Reynolds number are shown in Fig. 12. It can be seen that flow at lower Reynolds number (12.17 and 59.78) is qualitatively different than that at higher Reynolds numbers (204.74 and 2000). At higher Reynolds number, wake behind the spheres divides the high velocity stream in two parts at periodic planes (see two groups of red vectors in Fig. 12c and d). Profiles of predicted normalized z-component of the velocity on periodic plane at $x = 0.0015$ are shown in Fig. 13. At lower Reynolds number ($Re_p = 12.17$), flow profile is bell shaped with maximum velocity of about twice the mean velocity ($V_{max} = 2V_m$). With increase in Reynolds number, flow profile is flattened and wake behind the solid body starts affecting the flow profiles. At $Re_p = 2000$, splitting of high velocity stream into two parts is obvious from the shown velocity profiles (Fig. 13). Comparison of Figs. 13 and 9 clearly demonstrates the influence of packing arrangement on the flow in interstitial spaces. It can be seen from Fig. 9 that for SC arrangement the ratio of the z-component of velocity to the mean velocity is 2.5 for the lowest Reynolds number and increases up to 4 with increase in Reynolds number. For the rhombohedral arrangement, the value of this ratio is always below 2 (Fig. 13). Unlike SC, the flow encounters obstruction and changes the direction for rhombohedral arrangement of particles.

The predicted drag coefficients for FCC and SC geometry with solid volume fraction equal to 0.5 are shown in Fig. 14. It is quite clear that

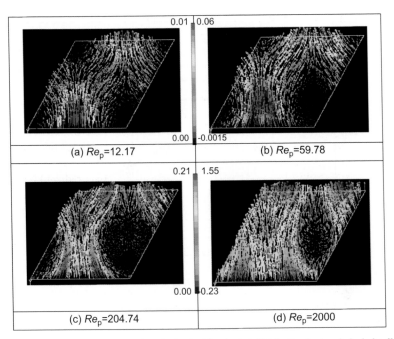

FIGURE 12 Velocity vectors and z-velocity (axial velocity) distribution in rhombohedral cell at various particle Reynolds numbers (Gunjal et al., 2005a). (a) $Re_p = 12.17$; (b) $Re_p = 59.78$; (c) $Re_p = 204.74$; (d) $Re_p = 2000$.

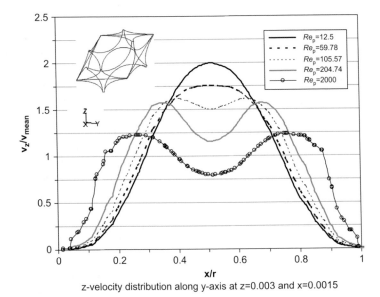

z-velocity distribution along y-axis at z=0.003 and x=0.0015

FIGURE 13 Velocity vectors and z-velocity (axial velocity) distribution in rhombohedral cell at various particle Reynolds numbers (Gunjal et al., 2005a).

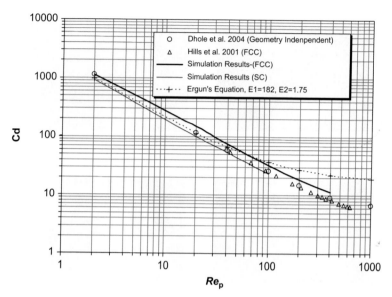

FIGURE 14   Comparison of simulated average drag coefficient with the literature data at various Reynolds numbers (Gunjal et al., 2005a).

geometrical orientation does not make a significant variation in overall drag force acting on the particle. However, the results of SC and one-dimensional rhombohedral geometry indicate that there is a significant difference in the predicted flow field distribution for these two cases. Figs. 9 and 13 also indicate that particle Reynolds number plays a significant role in the distribution of velocity inside the void space. Except the studies of Maier et al. (1998) and Magnico (2003) which report velocity distribution in void space at low Reynolds numbers, no other information is available in the literature on how particle arrangement and particle Reynolds number influence velocity distribution. The predicted distribution of z-velocity component is shown in Fig. 15 for one-dimensional rhombohedral geometry at different particle Reynolds numbers. At the lowest Reynolds number ($Re_p = 12.17$), z-velocity distribution exhibits a sharp peak and a shoulder. The predicted velocity distributions of rhombohedral arrangement are also similar to the results reported by Magnico (2003) and Maier et al. (1998). As the particle Reynolds number increases, the distribution broadens and becomes bi-modal. However, z-velocity distributions for three-dimensional rhombohedral geometry show different trends compared to the one-dimensional rhombohedral geometry (see Figs. 16 and 17). For three-dimensional rhombohedral geometry a flatter velocity distribution was observed at low as well as high Reynolds numbers.

Distributions of the axial component of velocity within the interstitial space for SC, rhombohedral (one-dimensional and three-dimensional), and FCC geometry were compared for Re_p 12.17 in Fig. 16. For SC and one-dimensional

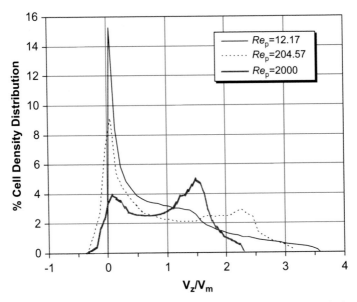

FIGURE 15 Histograms of z-velocity (axial velocity component) distribution obtained from simulations of liquid flow in one-dimensional rhombohedral cell at different Re_p (Gunjal et al., 2005a).

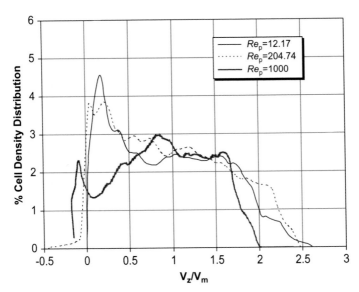

FIGURE 16 Histograms of z-velocity (axial velocity component) distribution obtained from simulations of liquid flow in three-dimensional rhombohedral cell at different Re_p (Gunjal et al., 2005a).

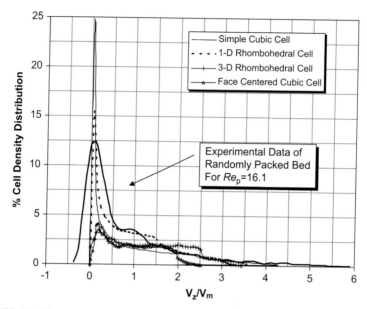

FIGURE 17 Histograms of z-velocity (axial velocity component) distribution obtained from simulations of liquid flow in different unit cells at $Re_p = 12.17$ and experimental data of Sederman et al. (1997) in randomly packed bed [$D = 4.7$ cm, $d_p = 5$ mm, normalized density distribution] (Gunjal et al., 2005a).

rhombohedral geometry, the predicted velocity distribution curves indicate that there are large numbers of cells containing low-magnitude axial velocities. However, for three-dimensional rhombohedral and FCC geometry a flatter velocity distribution was observed. Sederman et al. (1997) experimentally measured velocity distribution within the packed beds using MRI. Their data for the particle Reynolds number of 16.1 are also shown in Fig. 16. The velocity distribution obtained from experimental data of randomly packed bed lies between the trends observed in different arrangements considered here. Similar trends were observed in a predicted distribution of one of the transverse velocity components (x-velocity) in the interstitial space (see Fig. 18).

Total frictional resistance determined from the computational fluid dynamics (CFD) simulations showed good agreement with the values estimated using the Ergun equation (with $E_1 = 182$ and $E_2 = 1.75$) for all three packing arrangements. The Ergun equation was found to overpredict the drag coefficient for the turbulent flow regime. The ratio of surface drag to overall drag was almost independent of particle Reynolds number in the laminar flow regime. The values of this ratio obtained for the SC and FCC arrangements were almost the same (~0.21). For the rhombohedral arrangement (one-dimensional), the relative contribution of form drag was lower than that observed for the SC and FCC arrangements. The three-dimensional rhombohedral arrangement was

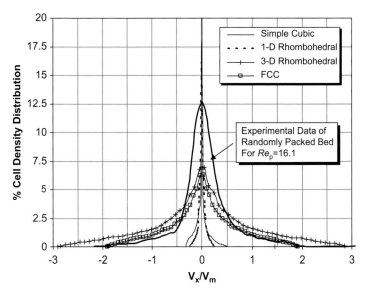

FIGURE 18   Histograms of x-velocity (transverse velocity component) distribution obtained from simulations of liquid flow in different unit cells at $Re_p = 12.17$ and experimental data of Sederman et al. (1997) in randomly packed bed ($D = 4.7$ cm, $d_p = 5$ mm, normalized density distribution) (Gunjal et al., 2005a).

found to offer maximum form drag. The predicted values of Nusselt numbers for the FCC arrangement showed reasonable agreement with the correlations of the particle to fluid heat transfer in packed beds. The predicted values of Nusselt number for the SC arrangement were much lower than those obtained for the FCC arrangement. Unlike the FCC, where flow impinges on the obstructing particle and changes directions several times within the "unit cell," no impingement and direction changes occur in the SC arrangement. This leads to significantly different path lines and velocity distribution and therefore the heat transfer. The unit cell approach can be used for quantitative understanding of influence of particle shapes and packing arrangement on flow and heat transfer.

Influence of Particle Shape on Flow Characteristics

Most of the results discussed so far were related to spherical particles. However, in practice several types of catalyst particles are used. For example, in cases where effectiveness factors are low, catalyst particles are "designed" to increase external surface area. Various types of catalysts have been designed with internal holes which increase surface area and bed porosity for improving performance (better catalyst activity and lower pressure drop). Some of these particle shapes are shown in Fig. 1 (of Chapter 1). Methodology of modeling of flow around an array of particles can be extended to particles of different shapes

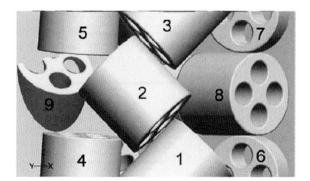

FIGURE 19 Assembly of cylindrical particles with holes considered by Nijemeisland, Dixon and Stitt (2004).

in a straightforward manner. As an example of this, a sample of results reported by Nijemeisland and Dixon (2004) is reproduced here. In their work, Nijemeisland and Dixon simulated flow and heat transfer over an assembly of particles of cylindrical shapes with internal holes (as shown in Fig. 19). Typical results are shown in Fig. 20. Path lines are colored by axial velocity in m/s. Dixon, Taskin, Nijemeisland, and Stit (2008) further extended this approach to quantify impact of particle shapes and packing on wall to particle heat transfer. It was observed that for comparable pressure drop costs, the multiholed particles lead to a lower tube wall temperature. The influence of changing diameter of holes for the four-small-holes and one-big-hole particles on flow and heat transfer was found to be rather small. The methodology and results discussed here can be used to guide design of catalyst particles for realizing better heat and mass transfer characteristics without incurring penalty of higher pressure drop. The approaches, models, and results discussed in this section are useful to gain better insight into flow within packed beds. The approaches and models used for simulating flow through larger packed beds are discussed in the following.

Flow Through a Packed Bed of Randomly Packed Particles

In the approach (A), as discussed earlier, instead of resolving individual particles in the packed bed, the entire packed bed region is treated as the porous media and flow through this media is simulated using averaged transport equations. This approach has been widely used to simulate flow through packed beds. For example, Ranade (1994) has used CFD model to optimize design of a deflector plate in an axial fixed-bed reactor. Foumeny and Benyahia (1993) also discussed application of CFD models for optimizing internals for axial flow fixed-bed reactors. Ranade (1997) has used this approach to improve the performance of radial flow packed bed reactor.

The key issue in modeling of fixed-bed reactor is the correct representation of a fixed bed (of solid particles) in a CFD model. In most of the cases, such

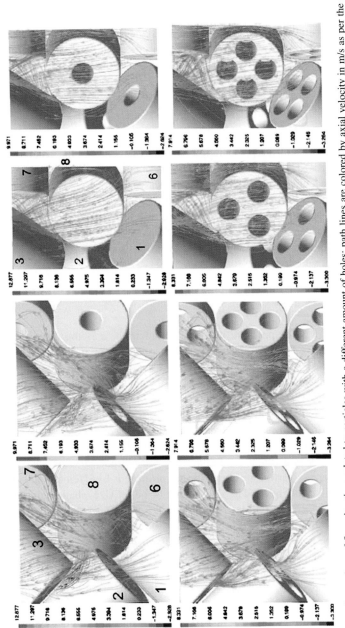

FIGURE 20 Comparison of flow situations related to particles with a different amount of holes; path lines are colored by axial velocity in m/s as per the scale shown on left side of the pictures (Nijemeisland, Dixon and Stitt, 2004).

a fixed bed of solid particles can be modeled as an isotropic or anisotropic porous media. Additional resistance offered by such porous media can then be modeled by introducing an additional momentum sink in the momentum transport equations. Usually additional resistance offered by porous media is represented in the flow model using two parameters, namely, permeability, α, and inertial coefficient, C_2. The additional momentum sink in terms of these two parameters is represented as:

$$\text{Momentum source} = -\left[\frac{\mu}{\alpha}V + C_2\left(\frac{1}{2}\rho V^2\right)\right] \tag{10}$$

Accuracy of such a representation obviously depends on the accuracy of the parameters used to represent the porous media. Best option to specify adequately accurate values of these parameters is, of course, experimental data. In the absence of such data, several correlations relating the pressure drop through porous beds and velocity and bed characteristics are available (see, for example, Carman, 1937; Ergun, 1952; Mehta & Hawley, 1969). The Ergun equation is one of the most widely used relations to represent the resistance of catalyst bed, which has the form:

$$\frac{\Delta P}{L} = \frac{E_1\mu}{d_p^2\phi_p^2}\frac{(1-\varepsilon)^2}{\varepsilon^3}V + \frac{E_2\rho}{d_p\phi_p}\frac{(1-\varepsilon)}{\varepsilon^3}V^2 \tag{11}$$

where $(\Delta P/L)$ is the pressure drop per unit length, μ is the viscosity, d_p is the equivalent pellet diameter, ϕ_p is the sphericity, ε is the porosity, and V is the superficial velocity. Where E_1 and E_2 are the Ergun's constants and are functions of bed characteristics. Ergun (1952) has recommended $E_1 = 150$ and $E_2 = 1.75$ for packed bed flow. Nemec and Levec (2005) have recently pointed out that these values of constants need to be modified to account for different particle shapes. The values of E_1 and E_2 increase with decrease in sphericity for cylindrical and polylobe particles. For estimating these constants as a function of sphericity for cylindrical, ring-shaped, and polylobe-shaped particles, correlations proposed by Nemec and Levec (2005) are recommended. The knowledge of pellet size, shape, and voidage of the bed are thus sufficient to characterize the resistance of the catalyst bed with the help of these correlations.

It must be noted that the porosity and its distribution in a packed bed are the key parameters in determining the flow distribution within the bed. Most of the correlations are usually obtained by considering parameters such as bed porosity averaged over the entire bed (ignoring spatial variation within the bed). Caution must be exercised in using such correlations for CFD simulations of flow through packed bed. In recent years, numerous attempts were made to provide quantitative information about the porosity distribution (see discussion in the section *Characterization of Packed Beds*). Mean porosity and its distribution are determined largely by particle size, shape, surface properties, and method of packing. These experimental and computational studies have shown that the cross-sectional-averaged porosity along the height of the bed is distributed randomly and the longitudinally averaged radial porosity profile exhibits a maximum near the wall. Bed porosity was also found to fluctuate significantly in the near wall region (of width of about 4−5 particle diameters). The magnitude of fluctuations is a strong function of a ratio of column diameter to particle diameter (for $D/d_p > 15$, fluctuations are within 1% whereas for lower values of D/d_p, fluctuations may rise up to 30%). The correlation of Mueller (1991) discussed in the section *Characterization of Packed Beds* is fairly general and is recommended for using with the CFD models for flow through packed beds.

CFD simulations with such randomly distributed bed porosity can provide useful insights and help evolve guidelines for improving flow distribution in packed beds. This approach can also be used as a basis for simulating more complex gas−liquid flow through packed bed (discussed in the following section).

GAS−LIQUID FLOW THROUGH PACKED BEDS

The presence of liquid in addition to the gas phase flowing through the packed bed offers several additional challenges in the analysis of the flow behavior compared to those mentioned in the previous section. As discussed previously, porosity is not uniform in randomly packed beds. This implies that some flow channels, formed within a packed bed, offer less resistance to flow than other channels. Liquid will tend to move toward channels of lower resistance, leading to higher liquid holdup in such channels. Thus, even if the initial liquid distribution is uniform, inherent random spatial variation of the bed leads to non-uniform liquid flow. Wetting of solid particles of the bed by liquid is significantly influenced by surface characteristics, bed structure, contact angle, capillary forces, and flow rates of liquid and gas streams. It is important to capture these complex interactions among gas, liquid, and solid phases adequately while developing a model for gas−liquid flow through packed beds.

The modeling approaches discussed above are still applicable in general. The meso-scale approaches are useful for gaining better insight into interaction of gas and liquid phases with solid particles, especially wetting. The macroscopic models are useful to provide overall understanding and behavior of the entire bed. The meso-scale models are expected to provide a better understanding of the flow structure around particles and closure models which might be useful for macro-scale models. It will therefore be useful to discuss macroscopic approaches/models before discussing meso-scale models. The basic macro-scale and meso-scale modeling approaches and model equations are discussed in the following sections. Applications of these models are discussed in the subsequent section.

Modeling of Gas–Liquid Flow Through Packed Bed

Macro-scale Models

Several different approaches such as percolation theory approach (Crine, Marchot, & L'Homme, 1979), network model (Thompson & Fogler, 1997), the Eulerian–Eulerian approach with the multifluid models (Attou & Ferschneider, 1999; Grosser, Carbonell, & Sundaresan, 1988; Jiang, 2000; Yin, Sun, Afacan, Nandakumar, & Chung, 2000), and the lattice Boltzmann-type models (Mantle et al., 2001) have been applied to simulate gas–liquid flow in packed beds. Attou and Ferschneider (1999) and Grosser et al. (1988) have used one-dimensional flow model with uniform porosity distribution in the bed. The Eulerian–Eulerian approach with the two-dimensional multifluid models with spatial variation of the porosity appears to be most suitable for reactor engineering applications (Gunjal et al., 2005; Jiang, Khadilkar, Al-Dahhan, & Dudukovic, 2002; Kashiwa, Padial, Rauenzahn, & Vander-Heyden, 1994; Ranade, 2002). In this approach, continuum approximation is applied for all the phases. Volume-averaged mass and momentum balance equations for the k-th fluid can be written as:

Mass balance equation:

$$\frac{\partial \varepsilon_k \rho_K}{\partial t} + \nabla \cdot \varepsilon_k \rho_k U_k = 0 \tag{12}$$

Momentum balance equation:

$$\frac{\partial (\varepsilon_k \rho_k U_k)}{\partial t} + \nabla \cdot \left(\varepsilon_k \rho_k U_k U_k \right) = -\varepsilon_k \nabla P_k + \nabla \cdot \left(\varepsilon_k \mu \nabla U \right) + \varepsilon_k \rho_k g$$
$$+ F_{K,R}(U_k - U_r) \tag{13}$$

where, ε_k represents the volume fraction of each phase, ρ_k is the density of k-th phase, U_k is the cell velocity of k-th phase, and P is a mean pressure shared by all the phases present in the system. $F_{K,R}$ is an interphase (between k and r phases) momentum exchange term. Left-hand side of Eq. (13) represents the rate of change of momentum for the k-th phase. The right-hand side represents, pressure forces, gravitational acceleration, average shear stresses, and interphase momentum exchange. It is convenient to treat gas phase as a primary phase and liquid–solid phases as secondary phases. Solid phase should be considered as stationary and distribution of solid volume fraction can be assigned at the beginning of the simulation following the methods discussed in the previous section.

The pressure drop in the packed bed is usually correlated using the Ergun equation or its variants (Al-Dahhan & Dudukovic, 1994; Holub, Dudukovic,

& Ramachandran, 1992; Saez & Carbonell, 1985). Interphase coupling terms may therefore be formulated based on similar equations. The presence of liquid flow, however, leads to additional interphase exchanges, which need to be formulated correctly. Different approaches viz., relative permeability model (Grosser et al., 1988; Saez & Carbonell, 1985), slit model (Holub, Dudukovic, & Ramachandran, 1993), and two-fluid interaction model (Attou & Ferschneider, 1999, 2000) have been proposed to formulate interphase momentum exchange terms. The model of Attou and Ferschneider (1999), which includes gas–liquid interaction force, has a theoretically sound basis and is discussed here.

The model of Attou and Ferschneider (1999) was developed for the regime in which liquid flows in the form of a film. In this work, we have explored the possibility of using this model for simulating flow regimes in which part of the liquid may flow in the form of droplets. The interphase coupling terms $F_{K,R}$ (given here by Eqs. 14–16) proposed by Attou and Ferschneider (2000) are rewritten in terms of interstitial velocities and phase volume fractions as (instead of superficial velocities and saturation):

Gas–liquid momentum exchange term:

$$
\begin{aligned}
F_{GL} = \varepsilon_G \Bigg(& \frac{E_1 \mu_G (1 - \varepsilon_G)^2}{\varepsilon_G^2 d_p^2} \left[\frac{\varepsilon_S}{(1 - \varepsilon_G)} \right]^{0.667} \\
& + \frac{E_2 \rho_G (U_G - U_L)(1 - \varepsilon_G)}{\varepsilon_G d_p} \left[\frac{\varepsilon_S}{(1 - \varepsilon_G)} \right]^{0.333} \Bigg)
\end{aligned}
\tag{14}
$$

Gas–solid momentum exchange term:

$$
\begin{aligned}
F_{GS} = \varepsilon_G \Bigg(& \frac{E_1 \mu_G (1 - \varepsilon_G)^2}{\varepsilon_G^2 d_p^2} \left[\frac{\varepsilon_S}{(1 - \varepsilon_G)} \right]^{0.667} \\
& + \frac{E_2 \rho_G U_G (1 - \varepsilon_G)}{\varepsilon_G d_p} \left[\frac{\varepsilon_S}{(1 - \varepsilon_G)} \right]^{0.333} \Bigg)
\end{aligned}
\tag{15}
$$

Liquid–solid momentum exchange term:

$$
F_{LS} = \varepsilon_L \left(\frac{E_1 \mu_L \varepsilon_S^2}{\varepsilon_L^2 d_p^2} + \frac{E_2 \rho_L U_G \varepsilon_S}{\varepsilon_L d_p} \right)
\tag{16}
$$

It must be noted that a pressure shared by all the phases is used in momentum balance equation (Eq. 13). However, when two immiscible phases are in contact with each other, interfacial tension causes the fluids to have different pressures. Such a pressure difference (capillary pressure) for gas and liquid phases may be written as:

$$P_G - P_L = 2\sigma\left(\frac{1}{d_1} - \frac{1}{d_2}\right) \tag{17}$$

where, d_1 and d_2 are the maximum and minimum diameters of the sphere with liquid film formed by the flowing liquid. More details of relating d_1 and d_2 to particle diameter, porosity, and the minimum equivalent diameter of the area between three particles in contact are given by Attou and Ferschneider (2000). Capillary pressure affects the liquid distribution and may set-up gradients of liquid holdup within the packed bed.

Several investigators have analyzed capillary forces (for example, Attou & Ferschneider, 2000; Grosser et al., 1988; Jiang et al., 2002; and references cited therein). Grosser et al. (1988) have studied onset of pulsing in trickle beds using linear stability analysis. Their analysis suggests that the competition between the inertial and capillary forces leads to a situation in which steady-state flow is not possible, implying the pulsing in trickle beds. Grosser et al. have proposed the capillary pressure as an empirical function of the liquid saturation:

$$P_L = P_G - \sigma\frac{\varepsilon_s E_1^{0.5}}{(1 - \varepsilon_s)d_p}[0.48 + 0.036\ln(1 - \beta_L)/\beta_L] \tag{18}$$

The order of magnitude analysis indicates that the magnitude of the capillary forces is rather small compared to the magnitudes of interphase drag forces. Attou and Ferschneider (2000) have obtained the following expression for the capillary pressure term based on geometric estimates of d_1 and d_2 and with empirical factor F to account for high pressure operations as:

$$P_G - P_L = 2\sigma\left(\frac{1 - \varepsilon}{1 - \varepsilon_G}\right)^{0.333}\left(\frac{5.416}{d_P}\right)F\left(\frac{\rho_G}{\rho_L}\right) \tag{19}$$

where,

$$F\left(\frac{\rho_G}{\rho_L}\right) = 1 + 88.1\frac{\rho_G}{\rho_L} \quad \text{for} \quad \frac{\rho_G}{\rho_L} < 0.025 \tag{20}$$

Under typical operating conditions of trickle beds, quantitative comparison of the capillary pressures estimated from Eqs. (18) and (19) is not very different (within 10%). Considering the geometric basis used by Attou and Ferschneider (1999), Eq. (19) is recommended for the macro-scale CFD model.

Pressure drop required to maintain specified gas and liquid throughputs is history dependent. Pressure drop at any specific liquid velocity measured with increasing liquid velocity is more than that measured with decreasing liquid velocity (see, for example, Szady & Sundaresan, 1991). Capillary phenomenon is one of the contributing factors of this observation. Jiang et al., (2002) have attempted simulation of this phenomenon by introducing an empirical factor (f) related to the degree of wetting in their capillary pressure formulation as:

$$P_G - P_L = (1 - f)P_c \tag{21}$$

For prewetted or fully wetted bed, f was set to one, implying zero capillary pressure. For non-wetted bed, f was set to zero (Jiang et al., 2002). For incorporating the capillary pressure in the CFD model, gradients of capillary pressure must be formulated as:

$$\frac{\partial P_G}{\partial z} - \frac{\partial P_L}{\partial z} = \frac{2}{3}\sigma \frac{5.416}{d_p} \left(\frac{\varepsilon_s}{1 - \varepsilon_G}\right)^{-2/3} \left(\left(\frac{1}{1 - \varepsilon_G}\right)\frac{\partial \varepsilon_s}{\partial z} \right.$$
$$\left. + \left(\frac{\varepsilon_s}{(1 - \varepsilon_G)^2}\right)\frac{\partial \varepsilon_G}{\partial z}\right)F\left(\frac{\rho_G}{\rho_L}\right) \tag{22}$$

Equation (22) can be used to express the gradients of liquid pressure (P_L) in the liquid phase momentum equations in terms of gradients of gas pressure (P_G) to incorporate the capillary pressure terms.

The model equations described above can be solved numerically with appropriate boundary conditions. Porosity distribution within the bed can be specified using the methods described while discussing single-phase flow through packed beds. At the inlet, velocity boundary conditions may be used. It is generally useful to minimize the discontinuity at the inlet. This can be accomplished by specifying the inlet boundary conditions based on estimated value of the overall liquid volume fraction, $\langle \varepsilon_L \rangle$ as:

$$U_{Lin} = \frac{(V_L/\rho_L)}{\langle \varepsilon_L \rangle} \qquad U_{Gin} = \frac{(V_G/\rho_G)}{1 - \langle \varepsilon_L \rangle} \tag{23}$$

where V_L and V_G are mass fluxes of liquid and gas phases (kg/m^{2}s), respectively, and ρ_L and ρ_G are densities of liquid and gas, respectively. No slip boundary condition can be used for all the impermeable walls.

Application of such macro-scale models for simulating trickle bed reactors involves extending these equations for considering species transport and chemical reactions. These extensions are discussed in a separate sub-section on *Simulation of Reactions in Trickle Bed Reactors*.

Meso-scale Models

Some of the important issues in macroscopic modeling of gas–liquid flow through packed beds are closure models for interphase drag, wetting, and capillary forces. It is essential to understand how liquid interacts with solid surfaces (either flat or curved) to make progress. One of the simplest "model" problem to understand such interactions is the interaction of liquid drops with flat and curved surfaces. The phenomenon of spreading of liquid drop on flat solid surface has been studied extensively (Gunjal, Chaudhari, & Ranade, 2005b and references cited therein).

Several methods are available to simulate free-surface flows (Fukai et al., 1995; Fukai, Zhao, Poulikakos, Megaridis, & Miyatake, 1993; McHyman, 1984; Monaghan, 1994; Ranade, 2002; Unverdi & Tryggvason, 1992) and motion of gas—liquid—solid contact line on solid surfaces. Free-surface methodologies can be classified into surface tracking, moving mesh, and fixed mesh (volume tracking) methods. Surface tracking methods define a sharp interface whose motion is followed using either a height function or marker particles. In moving mesh methods, a set of nodal points of the computational mesh is associated with the interface. The computational grid nodes are moved by interface fitted mesh method or by following the fluid. Both of these methods retain the sharper interface. However, mesh or marker particles have to be relocated and re-meshed when interface undergoes large deformations. As the free-surface deformation becomes complex, the application of these methods becomes computationally very intensive. Another method, which can retain a sharp interface, is the boundary integral method (Davidson, 2000, 2002). However, use of this method is still mainly restricted to two-dimensional simulations.

Volume of the fluid (VOF) method developed by Hirt and Nichols (1981) is one of the most widely used methods in modeling of free surfaces. This is a fixed mesh method, in which, the interface between immiscible fluids is modeled as the discontinuity in characteristic function (such as volume fraction). Several methods are available for interface reconstruction such as SLIC (simple line interface calculation), PLIC (piece-wise linear interface calculations), and Young's PLIC method with varying degree of interface smearing (Ranade, 2002; Rider & Kothe, 1995; Rudman, 1997; for more details). The governing equations and other associated closure models may be found elsewhere (for example, Gunjal et al., 2005b; Lopes & Quinta-Ferreira, 2009; Ranade, 2002; and references cited therein). This approach has been used for quantitative understanding of spreading of liquid film and interaction of droplets with flat and curved surfaces (see, for example, Gunjal et al., 2005b).

The flow field predicted by the VOF models can be used to examine various intricate details of interaction of a liquid drop and flat surface. Gunjal et al. (2005b) have demonstrated this by critically analyzing variation of kinetic, potential, and surface energies during drop impact, spreading, and oscillation processes. When a drop spreads to its maximum extent and is about to recoil, its potential energy exhibits minimum. For every cycle of potential energy, there are two cycles of kinetic energy because it passes through maximum during spreading as well as recoiling. Scales of variation

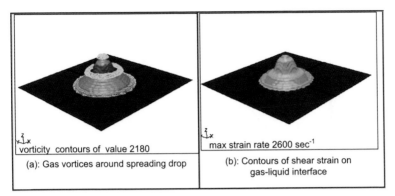

vorticity contours of value 2180

(a): Gas vortices around spreading drop

max strain rate 2600 sec^{-1}

(b): Contours of shear strain on
gas-liquid interface

FIGURE 21   Illustration of gas–liquid and liquid–solid interactions during drop spreading on glass surface at time = 7.5 ms (Gunjal, Chaudhari, & Ranade, 2005b). System: 3 mm water drop with impact velocity 0.3 m/s on glass surface of static contact angle 64°. (a) Gas vortices around spreading drop; (b) contours of shear strain on gas–liquid interface.

of surface energy are higher than the potential and kinetic energy and its variation is very sensitive to the variation in a contact angle. Therefore, small errors in the values of contact angle or surface area may corrupt the calculation of total energy. The detailed flow field predicted by such VOF simulations can be used to compute other quantities of interest such as gas–liquid and liquid–solid interactions (as shown in Fig. 21). Liquid–solid interaction can be determined by calculating the average shear stress exerted by the fluid on the solid surface (see Fig. 21). Gas–liquid interaction can be studied by calculating the strain rate on gas–liquid interface. Detailed study of gas–liquid interaction (in terms of strain rate), gas recirculation (in terms of vorticity) as illustrated in Fig. 21 will be useful for understanding interphase heat, mass, and momentum transfer in such flows. Maximum drop interfacial area was observed when the drop spreads completely; strain rate is smallest at this point. Microscopic evaluation of these parameters would be eventually useful for developing better closure terms for macro-scale flow models.

Recently Lopes and Quinta-Ferreira (2009) have used VOF approach to simulate gas–liquid flow through an array of solid particles. The computational model was used to quantify effect of gas and liquid flow rates on frictional pressure drop and liquid holdup. Some of the results obtained by them are reproduced in Figs. 22–26. The meso-scale models with VOF approach capture complex interactions of gas–liquid–solid phases reasonably well (at least for low gas velocities).

These meso-scale models can in principle provide useful quantitative information for developing closure models for the macro-scale simulations. These models can also help extrapolating our understanding and data for higher temperatures and pressures relevant to industrial trickle bed reactors.

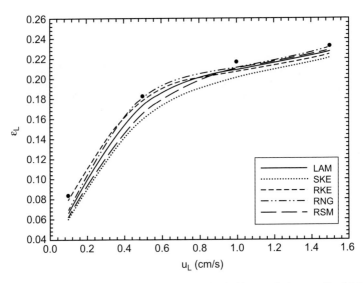

FIGURE 22 Influence of turbulence model on liquid holdup predictions at $G = 0.1 \text{ kg/m}^2\text{s}$ (Lopes & Quinta-Ferreira, 2009).

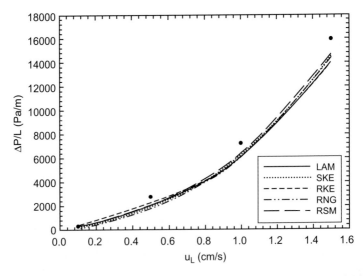

FIGURE 23 Influence of turbulence model on pressure drop predictions at $G = 0.1 \text{ kg/m}^2\text{s}$ (Lopes & Quinta-Ferreira, 2009).

Further work using this approach is needed for creating a comprehensive basis for selecting appropriate closure models for the macro-scale simulations. A sample of results obtained with the macro-scale models is discussed in the following.

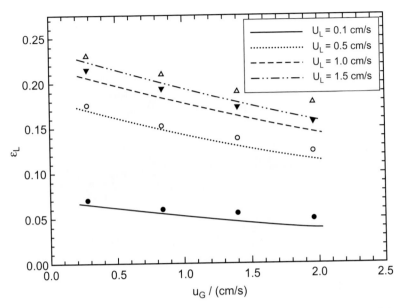

FIGURE 24 Influence of gas and liquid velocities on liquid holdup predictions (Lopes & Quinta-Ferreira, 2009).

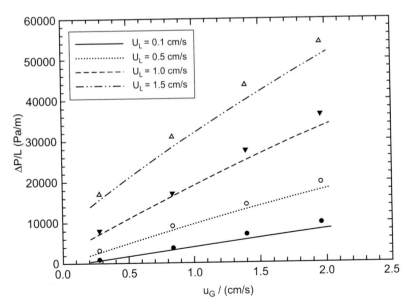

FIGURE 25 Influence of gas and liquid velocities on pressure drop predictions (Lopes & Quinta-Ferreira, 2009).

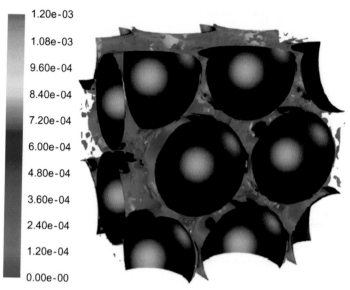

1.20e-03

1.08e-03

9.60e-04

8.40e-04

7.20e-04

6.00e-04

4.80e-04

3.60e-04

2.40e-04

1.20e-04

0.00e-00

FIGURE 26 Instantaneous snapshot of liquid holdup isosurface (el = 0.15) colored by liquid velocity magnitude [$L = 5$ kg/m^{2}s, $G = 0.7$ kg/m^{2}s, $P = 30$ bar, $d_p = 2$ mm] (Lopes & Quinta-Ferreira, 2009).

Simulation of Gas–Liquid Flow in Trickle Beds

Recently, Gunjal et al. (2005) reported experimental data and CFD results on trickle beds of two different diameters with two particle sizes. Some of their results are reproduced here to illustrate the application of CFD model for simulating macro-scale flow characteristics in a trickle bed. The CFD model was then used to estimate the extent of suspended liquid in trickle flow regime and to simulate periodic operation of trickle beds to seek an insight into the pulse flow regime.

As discussed earlier, measurements of pressure drop and liquid holdup for trickle bed reactor showed hysteresis with liquid velocity. The observed hysteresis is associated with the capillary pressure acting on three-phase contact line. Accurate representation of capillary term in CFD model is difficult. The simulations were carried out by setting value of "f" from Eq. (21) to one for prewetted beds and to zero for a dry bed. The simulated results are compared with the experimental data in Figs. 27 and 28. As observed in the experiments, simulated results showed lower pressure drop for the dry bed compared to the prewetted bed. The predicted magnitude of the hysteresis is, however, lower than that observed in the experiments. For the dry bed (lower branch as well as the initial part of the upper branch), experimental data showed non-linear variation of pressure drop with liquid velocity. However, simulated results showed almost a linear variation. The inadequate representation of

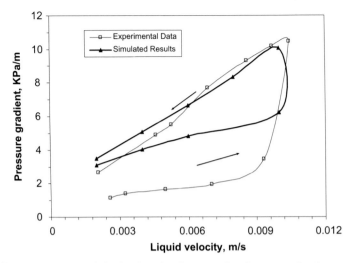

FIGURE 27 Comparison of simulated results of pressure drop for prewetted and non-prewetted beds [$V_G = 0.22$ m/s, std dev $= 5\%$, $D = 0.114$ m, $d_p = 3$ mm] (Gunjal et al., 2005).

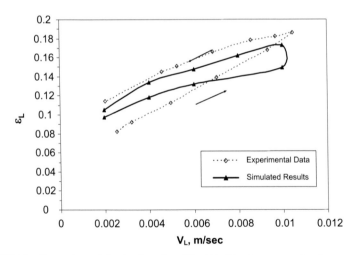

FIGURE 28 Comparison of simulated results of liquid holdup for prewetted and non-prewetted beds [$V_G = 0.22$ m/s, std dev $= 5\%$, $D = 0.114$ m, $d_p = 3$ mm] (Gunjal et al., 2005).

capillary forces is the most likely cause of this discrepancy. Non-linearity appears even in upper branch mainly because, when liquid velocity is reduced, partial dryout may occur in the bed making it similar to the non-prewetted bed. Simulated contours of liquid holdup for prewetted bed and non-prewetted bed are shown in Fig. 29. Liquid distribution in prewetted bed is relatively uniform as compared to the non-prewetted bed conditions (Fig. 29). It can be seen from

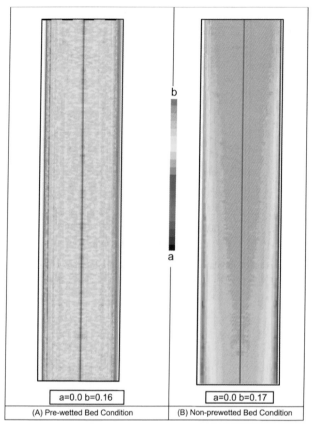

| a=0.0 b=0.16 | a=0.0 b=0.17 |
| (A) Pre-wetted Bed Condition | (B) Non-prewetted Bed Condition |

FIGURE 29 Simulated contours of liquid holdup for prewetted and non-prewetted bed [$V_L = 6$ kg/m^{2}s, $V_G = 0.22$ m/s, std dev = 5%, $D = 0.114$ m, $d_p = 3$ mm] (Gunjal et al., 2005). (A) Prewetted bed condition; (B) non-prewetted bed condition.

Fig. 30 that velocity and holdup distributions within the bed for liquid phase for prewetted and non-prewetted beds are substantially different. Distributions for the prewetted bed are wider than non-prewetted bed.

For prewetted bed, simulated results of pressure drop for the 0.114 m diameter of column are compared with the experimental data (Fig. 31) for gas velocities 0.22 m/s and 0.44 m/s. Simulated results overpredict pressure drop for 3 mm particle at low gas velocity (0.22 m/s) and underpredict for high gas velocity ($V_G = 0.44$ m/s). However, for 6 mm particles, simulated results show reasonable agreement with the experimental data. The disagreement at low liquid velocity for $V_G = 0.44$ m/s may be due to dryout phenomenon at high gas flow rate. Influence of column diameter on the agreement between simulated and experimental data is shown in Fig. 32. It can be seen that the agreement is better for 6 mm particles than for 3 mm particles. One of the possible reasons

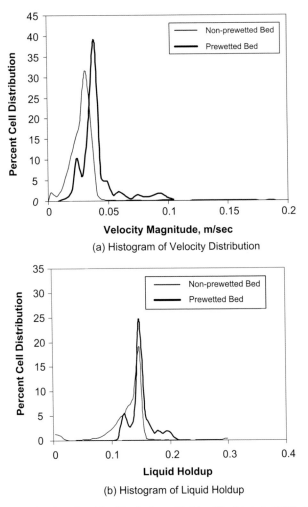

(a) Histogram of Velocity Distribution

(b) Histogram of Liquid Holdup

FIGURE 30 Simulated results of liquid velocity and holdup (Gunjal et al., 2005). (a) Histogram of velocity distribution; (b) histogram of liquid holdup.

for this could be inadequacies in representation of appropriate porosity distribution for 3 mm particle. Simulated values of the total liquid holdup in column of 0.114 m diameter are compared with the experimental data in Fig. 33. Simulated results show good qualitative as well as quantitative agreement with the experimental data. Similar agreement was also found for liquid holdup in 0.194 m diameter column (see Fig. 34). The simulated results can also be used to gain an insight in to hydrodynamics of gas–liquid flow through trickle beds.

In order to illustrate the application over a wider range of operating conditions, the simulations of three independent experimental data sets carried

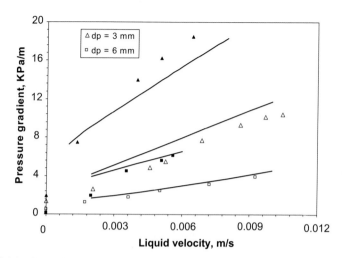

FIGURE 31 Comparison of CFD results with experimental data (Gunjal et al., 2005). Filled symbols: $V_G = 0.44$ m/s. Unfilled symbols: $V_G = 0.22$ m/s, $D = 0.114$ m.

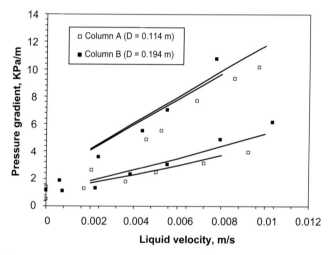

FIGURE 32 Comparison of CFD results with experimental data with $d_p = 3$ mm (Gunjal et al., 2005). Filled symbols: $V_G = 0.44$ m/s. Unfilled symbols: $V_G = 0.22$ m/s.

out by Gunjal et al. (2005) are briefly discussed here. Details of bed characteristics and operating conditions used in these three cases are summarized in Table 1. In Case 1, column to particle diameter (D/d_p) ratio was much higher compared to the other two cases. In Case 1 variation of pressure drop and liquid saturation was studied with change in liquid mass flow rate, while effect of change in gas mass flow rate on hydrodynamic parameters was studied in the other two cases. Gas mass flow rate in Case 3 was much higher than the other

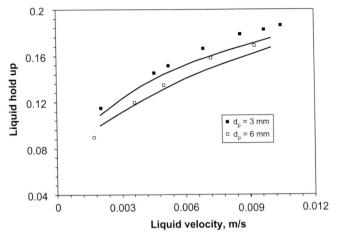

FIGURE 33 Comparison of CFD results with experimental data (Gunjal et al., 2005) [$D = 0.114$ m, $V_G = 0.22$ m/s].

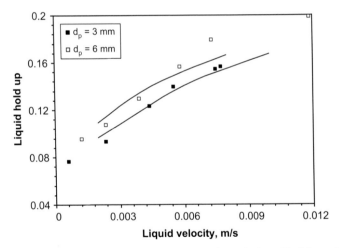

FIGURE 34 Comparison of CFD results with experimental data (Gunjal et al., 2005) [$D = 0.194$ m, $V_G = 0.22$ m/s].

two cases. Comparison of the experimental data with the simulated results is discussed in the following section.

Comparison of simulated results with the experimental data of Szady and Sundaresan (1991) for pressure drop and total liquid saturation is shown in Fig. 35a (only upper branch of pressure drop curve is shown here). The predicted results showed the correct trends of variation of the pressure drop and liquid saturation with liquid mass flow rate. The values of Ergun's (E_1 and E_2)

TABLE 1 Details of Bed Characteristics and Operating Conditions Used in Simulations

Case	Data Source	D/d_p	Bed Characteristics	V_G (m/s)	$V_L \times 10^3$ (m/s)	Ergun's Constants (E_1 and E_2)
1	Szady and Sundaresan (1991)	55	3 mm spherical; $\varepsilon_B = 0.37$	0.22	0.2−0.8	215 and 1.75
2	Specchia and Baldi (1977)	29.6	2.7 mm spherical; $\varepsilon_B = 0.38$	0.2−0.8	2.8	500 and 3
3	Rao et al. (1983)	15.4	3 mm spherical; $\varepsilon_B = 0.37$	1.5−5.5	1.0	215 and 3.4

constants used in the closure models represent the bed packing characteristics. In this case, $E_1 = 215$ and $E_2 = 1.75$ were used for carrying out the simulations. At low liquid velocities, the model showed good agreement with the experimental data. At higher liquid velocities (~8 kg/m²s), model underpredicted the pressure drop values. This may be because of the possible transition from trickle flow regime to pulse flow regime. Model predictions for pressure drop and liquid saturation for experimental data of Specchia and Baldi (1977) is shown in Fig. 35b. In this case, higher values of Ergun's constants ($E_1 = 500$ and $E_2 = 3$) were used in the model. These values are close to the values suggested by Holub et al. (1992). In this case, model predictions showed good agreement with the experimental data of pressure drop and liquid saturation. A third set of data from Rao, Ananth, and Varma (1983), which was obtained for very high gas velocities compared to earlier data sets (1−8 kg/m²s), was also used for evaluating the model predictions. As velocity of the gas phase increases (>1.5 kg/m²s), trickle flow regime may change to spray flow regime. It can be seen from Fig. 35c that the model predictions showed good agreement with the experimental data for the high gas velocity cases as well (with $E_1 = 215$ and $E_2 = 3.4$). Though, the value of E_2 looks rather high, it is well within the range of values used by previous investigators (see a review given by Holub et al., 1992).

The simulated results show reasonable agreement with the experimental data of various studies. The basic CFD model developed in this work uses two parameters (E_1 and E_2) to match the predicted pressure drop with the reported experimental data. Reported results indicate that with appropriate values of these parameters, CFD model is able to predict the overall liquid saturation

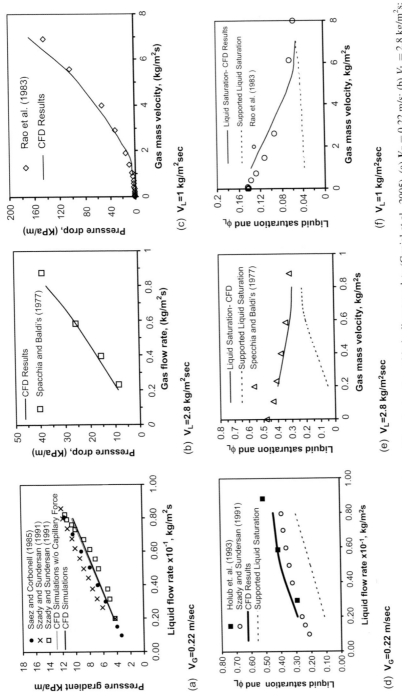

FIGURE 35 Comparison of simulated pressure drop and liquid saturation with the literature data (Gunjal et al., 2005). (a) $V_G = 0.22$ m/s; (b) $V_L = 2.8$ kg/m^{2}s; (c) $V_L = 1$ kg/m^{2}s; (d) $V_G = 0.22$ m/s; (e) $V_L = 2.8$ kg/m^{2}s; (f) $V_L = 1$ kg/m^{2}s. Other details of three experimental sets are listed in Table 1.

correctly for the range of gas and liquid velocities. Encouraged by such an agreement, the CFD model developed here was used to estimate frictional pressure drop and supported liquid saturation. For this purpose, simulations were carried out for two values of gravitational acceleration. For the detailed procedure to calculate the frictional pressure drop and supported liquid holdup, the reader is referred to the paper by Gunjal et al. (2005).

Following their procedure, frictional pressure drop and liquid saturation supported by gas phase were calculated. The predicted values of supported liquid saturation are shown by a dotted line in Fig. 35d–f. Figure 35d shows that fraction of liquid supported by gas phase increases with liquid velocity. The results shown in Fig. 35e and f indicate that the fraction of supported liquid saturation increases with gas velocity though increase in gas velocity decreases total liquid holdup or saturation. At very high gas velocity (see Fig. 35f) almost all the liquid holdup is supported by gas phase, indicating a spray flow regime. Thus, the procedure of Gunjal et al. (2005) can be used to estimate fraction of liquid supported by the gas phase and to identify possible transition to a spray flow regime.

In many industrial practices, liquid-induced periodic operations are preferable because reactor operation with natural pulsing is difficult for large reactors. It is worthwhile to apply present CFD model to simulate periodic operation of a trickle bed. This will also help to gain an insight into the features of the pulse flow regime. In natural pulse flow regime, liquid-enriched pulses form after some distance from the inlet and they accelerate while moving downward. Formation of pulses is associated with complex interactions among capillary forces, wall adhesion, and the convective forces. If the model equations adequately represent this underlying physics and numerical solution does not add any artificial diffusivity, the simulated results should be able to capture transition from trickle flow regime to pulse flow regime. However, the current understanding of physics of pulse formation and its implementation in CFD model is not adequate for this purpose. Alternative way to gain some insight into pulse-like flow in trickle beds is to simulate periodic operation of trickle beds with induced pulses by manipulating inlet liquid velocity. Boelhouwer, Piepers, and Drinkenburg (2002) have compared the key features of natural pulsing and induced pulsing trickle beds. Their results showed that the variation between natural pulsing and induced pulsing is within 25%. Following this, Gunjal et al. (2005) had simulated flow in trickle beds operated with liquid-induced pulsing maintaining the same average flow rates. The frequency of liquid-induced pulsing was set from the experimental measurements of natural pulsing. They carried out simulations of liquid-induced periodic operation for different liquid flow rates ($V_L = 11-24$ kg/m^{2}s) at a gas flow rate of $V_G = 0.22$ m/s. The predicted pressure drop was not found to be sensitive to the pulsing frequency within the range studied. The models discussed in this section can, however, be used to simulate periodic operation of trickle bed reactor as well.

Simulation of Reactions in Trickle Bed Reactors

CFD models can be used to simulate liquid phase mixing and reactions in trickle bed reactors. Ability to represent porosity variation within the bed and to consider detailed three-dimensional configurations of real-life reactors offers new possibilities of gaining insight and of improving performance of such industrial reactors. For extending the CFD models described earlier to simulate liquid phase mixing and reactions, species mass balance equations need to be solved in addition to the overall mass, momentum, and energy conservation equations. These additional species conservation equations are of the form:

$$\frac{\partial \varepsilon_k \rho_k C_{k,i}}{\partial t} + \nabla \cdot \left(\varepsilon_k \rho_k U_k C_{k,i} \right) = \nabla \left(\varepsilon_k \rho_k D_{i,m} \nabla C_{k,i} \right) + \varepsilon_k \rho_k S_{k,i} \quad (24)$$

where, $C_{k,i}$ is the concentration of species i in k-th phase (gas or liquid) and ρ_k and ε_k are the density and volume fraction of the k-th phase, respectively. $S_{k,i}$ is the source for species i in phase k.

The formulation of source terms depends on gas—liquid, gas—solid, and liquid—solid mass transfer, extent of wetting of particles, extent of intraparticle concentration profiles, and reaction kinetics. The source term for species i in gas phase can be written as:

$$S_i = -k_{GLi}a_{GL} \left[\frac{C_{Gi}}{H_i} - C_{Li} \right] - k_{GSi}a_{GS}[C_{Gi} - C_{Si}] \quad (25)$$

where "k" denotes mass transfer coefficient and "a" denotes interfacial area per unit volume of the reactor. The subscripts GL and GS denote gas to liquid and gas to solid, respectively. It should be noted that a_{GS} will be zero if particles are completely wetted (that is, when gas phase is not directly in contact with the solid particles). The corresponding source term for species i in liquid phase may be written as,

$$S_i = k_{GLi}a_{GL} \left[\frac{C_{Gi}}{H_i} - C_{Li} \right] - k_{LSi}a_{LS}[C_{Li} - C_{Si}] \quad (26)$$

where subscript LS denotes liquid to solid. The unknown solid concentrations are calculated by assuming no accumulation on solid as:

$$k_{GSi}a_{GS}[C_{Gi} - C_{Si}] + k_{LSi}a_{LS}[C_{Li} - C_{Si}] = \rho_B \sum_{j=1}^{j=nr} \eta_{oj} r_{ij} \quad (27)$$

where, r is the rate of reaction in consistent units (kg/kg of catalysts), nr is the number of reactions, and ρ_B is the catalyst bulk density of the bed, kg of catalyst/m^3 of bed. η_{oj} is the overall catalyst particle effectiveness factor (weighted average of wetted and unwetted portion) for reaction j.

Physico-chemical properties of the systems (like density, viscosity, heat capacity, etc.) will in general be functions of species concentration and temperature. Appropriate correlations to estimate these properties can be incorporated in the CFD framework following standard practices (see Ranade, 2002 for more details). For interphase mass and heat transfer, appropriate correlations need to be selected (see Chapter 2 for discussion on various correlations which may be used for this purpose). With some of these empirical correlations, CFD models allow simulation of various key parameters like liquid holdup and its distribution within the bed and RTD without any further empiricism. This avoids the need of approximating liquid phase mixing (use of correlations to estimate dispersion coefficient) and liquid holdup/distribution. These types of CFD models have been shown to capture liquid phase mixing and reactions in trickle bed reactor reasonably well. For example, Gunjal, Ranade, and Chaudhari (2003) simulated liquid phase RTD using a similar model. Comparison of predicted RTD with the experimental data is shown in Fig. 36. Predicted RTD response is reasonably close to the experimental values. The CFD models can capture influence of porosity variation within the bed as well as key design issues and hardware details like location and sizing of inlet nozzles, supports, and distributors on hydrodynamics and therefore on reactor performance. Application of the models discussed here for simulating labora-tory-scale and commercial-scale hydroprocessing reactors is discussed in Chapter 6. Application of Chapters 2–4 for reactor performance evaluation and scale-up is discussed in Chapter 5.

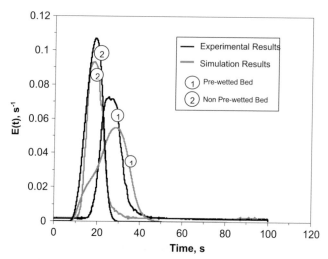

FIGURE 36 Comparison of the predicted residence time distribution with the experimental data (Gunjal et al., 2003). System: $D = 0.114$ m, $d_p = 6$ mm, $V_L = 1.72$ kg/m^{2}s, $V_G = 0.22$ m/s, std. dev. $= 5\%$, $d_p = 6$ mm, $E_1 = 500$, and $E_2 = 3.4$.

SUMMARY

Ability to predict and control the underlying fluid dynamics is essential to gain an insight for developing innovative engineering solutions. With the current level of understanding, it is more advantageous to use hierarchy of models coupled with key experimental findings to achieve practical benefits. In this chapter, therefore, different modeling approaches which allow better quantitative understanding of the processes occurring in trickle bed reactors are discussed. Different approaches for simulating single-phase flow through packed bed were first discussed. A "unit cell" approach or considering an assembly of few particles is very useful for obtaining quantitative insights into flow structures and transport processes occurring in packed beds. This approach can also be extended to gas–liquid flow through packed beds using VOF (volume of fluid) models. Information obtained from these models can be used for developing appropriate closures for macroscopic models. The Eulerian–Eulerian approach is recommended for modeling of macro-scale flow processes in fixed bed as well as trickle bed reactors. The realistic representation of the characteristics of the fixed bed (porosity distribution, degree of anisotropy, and so on) is crucial for carrying out simulations for engineering applications. Most of the current work relies on empirical information and pressure drop data to calibrate computational flow models of fixed and trickle bed reactors. Such calibrated computational flow models will be useful for understanding issues related to maldistribution, channeling, formation of hotspots, etc. To realize process intensification and performance enhancement, accurate knowledge of the underlying flow field in chemical reactors is essential. The approach and computational models developed in this book will allow reactor engineer to harness the power of computational flow modeling for better reactor engineering.

Adequate attention to some of the key issues mentioned here and creative use of computational flow modeling will make significant contributions in enhancing engineering insight and engineering practice. Applications of various modeling aspects discussed so far to reactor performance evaluation and scale-up with some examples are discussed in the following two chapters.

REFERENCES

Al-Dahhan, M. H., & Dudukovic, M. P. (1994). Pressure drop and liquid holdup in high pressure trickle-bed reactors. *Chemical Engineering Science, 49*(24), 5681.

Al-Dahhan, M. H., Wu, Y., & Dudukovic, M. P. (1995). Reproducible technique for packing laboratory-scale trickle-bed reactors with a mixture of catalyst and fines. *Industrial and. Engineering Chemistry Research, 34*(3), 741−747.

Attou, A., & Ferschneider, G. A. (1999). Two-fluid model for flow regime transition in gas–liquid trickle-bed reactors. *Chemical Engineering Science, 54*(21), 5031.

Attou, A., & Ferschneider, G. (2000). A two-fluid hydrodynamic model for the transition between trickle and pulse flow in a cocurrent gas–liquid packed-bed reactor. *Chemical Engineering Science, 55*, 491.

Boelhouwer, J. G., Piepers, H. W., & Drinkenburg, A. A. H. (2002). Liquid-induced pulsing flow in trickle-bed reactors. *Chemical Engineering Science, 57,* 3387.

Calis, H. P. A., Nijenhuis, J., Paikert, B. C., Dautzenberg, F. M., & van den Bleek, C. M. (2001). CFD modelling and experimental validation of pressure drop and flow profile in a novel structured catalytic reactor packing. *Chemical Engineering Science, 56,* 1713.

Carman, P. C. (1937). Fluid flow through granular beds. *Trans. Inst. Chem. Eng., 15,* 150—166.

Crine, M., Marchot, P., & L'Homme, G. A. (1979). Mathematical modeling of the liquid trickling flow through a packed bed using the percolation theory. *Computers and Chemical Engineering, 3,* 515.

Davidson, M. R. (2000). Boundary integral prediction of the spreading of an inviscid drop impacting on a solid surface. *Chemical Engineering Science, 55,* 1159.

Davidson, M. R. (2002). Spreading of inviscid drop impacting on liquid film. *Chemical Engineering Science, 57,* 3639.

Dhole, S. D., Chhabra, R. P., & Eswaran, V. (2004). Power law fluid through beds of spheres at intermediate Reynolds numbers: pressure drop in fixed and distended beds. *Chemical Engineering Research and Design, 82*(A6), 1—11.

Dixon, A. G., & Nijemeisland, M. (2001). CFD as a design tool for fixed-bed reactors. *Industrial and Engineering Chemistry Research, 40,* 5246.

Dixon, A. G., Taskin, M. E., Nijemeisland, M., & Stit, E. H. (2008). Wall-to-particle heat transfer in steam reformer tubes: CFD comparison of catalyst particles. *Chemical Engineering Science, 63*(8), 2219—2224.

Donohue, T. J., & Wensrich, C. M. (2008). A numerical investigation of the void structure of fibrous materials. *Powder Technology, 186*(1), 72—79.

Durst, F., Haas, R., & Interthal, W. (1987). The nature of flows through porous media. *Journal of Non-Newtonian Fluid Mechanics, 22,* 169.

Ergun, S. (1952). Flow through packed columns. *Chemical Engineering Progress, 48,* 89—94.

Foumeny, E. A., & Benyahia, F. (1993). Can CFD improve the handling of air, gas and gas—liquid mixtures? *Chemical Engineering Progress, 91*(1), 8—9, 21.

Freund, H., Zeiser, T., Huber, F., Klemm, E., Brenner, G., Durst, F., et al. (2003). Numerical simulations of single phase reacting flows in randomly packed/fixed-bed reactors and experimental validation. *Chemical Engineering Science, 58,* 903.

Fukai, J., Shiiba, Y., Yamamoto, T., Miyatake, O., Poulikakos, D., Megaridis, C. M., et al. (1995). Wetting effects on the spreading of a liquid droplet colliding with a flat surface: experiment and modeling. *Physics of Fluids, 7,* 236.

Fukai, J., Zhao, Z., Poulikakos, D., Megaridis, C. M., & Miyatake, O. (1993). Modeling of the deformation of a liquid droplet impinging upon at surface. *Physics of Fluids A, 5,* 2588.

Grosser, K., Carbonell, R. G., & Sundaresan, S. (1988). Onset of pulsing in two-phase concurrent downflow through a packed bed. *AIChE Journal, 34,* 185.

Gunjal, P. R., & Ranade, V. V. (2007). Modeling of laboratory and commercial scale hydro-processing reactors using CFD. *Chemical Engineering Science, 62,* 5512—5526.

Gunjal, P. R., Ranade, V. V., & Chaudhari, R. V. (2003). Liquid phase residence time distribution in trickle bed reactors: experiments and CFD simulations. *The Canadian Journal of Chemical Engineering, 81,* 821.

Gunjal, P. R., Kashid, M. N., Ranade, V. V., & Chaudhari, R. V. (2005). Hydrodynamics of trickle-bed reactors: experiments and CFD modeling. *Industrial and Engineering Chemistry Research, 44,* 6278—6294.

Gunjal, P. R., Chaudhari, R. V., & Ranade, V. V. (2005a). Computational study of a single-phase flow in packed beds of spheres. *AIChE Journal, 51*(2), 365—378.

Gunjal, P. R., Chaudhari, R. V., & Ranade, V. V. (2005b). Dynamics of drop impact on solid surface: experiments and VOF simulations. *AIChE Journal, 51*(1), 59—78.

Hill, R. J., Koch, D. L., & Ladd, A. J. C. (2001). Moderate-Reynolds-number flows in ordered and random arrays of spheres. *Journal of Fluid Mechanics, 448*, 243.

Hirt, C. W., & Nichols, B. D. (1981). Volume of fluid (VOF) method for the dynamics of free boundaries. *Journal of Computational Physics, 39*, 201.

Holub, R. A., Dudukovic, M. P., & Ramachandran, P. A. (1992). A phenomenological model for pressure drop, liquid holdup, and flow regime transition in gas—liquid trickle flow. *Chemical Engineering Science, 47*(9—11), 2343.

Holub, R. A., Dudukovic, M. P., & Ramachandran, P. A. (1993). Pressure drop, liquid hold-up and flow regime transition in trickle flow. *AIChE Journal, 39*, 302.

Irandoust, S., & Anderson, B. (1988). Monolithic catalysts for nonautomobile applications. *Catalysis Reviews: Science and Engineering, 30*(3), 341—392.

Jiang, Y. (2000). Flow Distribution and its Impact on Performance of Packed-Bed Reactors. *Ph.D. Thesis*. St. Louis, MO: Washington University.

Jiang, Y., Khadilkar, M. R., Al-Dahhan, M. H., & Dudukovic, M. P. (2000). Single phase flow modeling in packed beds: discrete cell approach revisited. *Chemical Engineering Science, 55*(10), 1829—1844.

Jiang, Y., Khadilkar, M. R., Al-Dahhan, M. H., & Dudukovic, M. P. (2001). CFD modeling of multiphase flow distribution in catalytic packed bed reactors: scale down issues. *Catalysis Today, 66*, 209—218.

Jiang, Y., Khadilkar, M. R., Al-Dahhan, M. H., & Dudukovic, M. P. (2002). CFD modeling of multiphase in packed bed reactors: results and applications. *AIChE Journal, 48*, 716.

Jongen, T. (1992). In: *Simulation and Modeling of Turbulent Incompressible Flows. PhD Thesis*. EPF Lausanne. Lausanne, Switzerland.

Kader, B. (1993). Temperature and concentration profiles in fully turbulent boundary layers. *International Journal of Heat Mass Transfer, 24*(9), 1541.

Kashiwa, B. A., Padial, N. T., Rauenzahn, R. M., & Vander-Heyden, W.B. (1994). A cell centered ICE method for multiphase flow simulations. In ASME symposium on numerical methods for multiphase flows. Lake Tahoe, NV.

Kim, S. E., & Choudhury, D. (1995). A near-wall treatment using wall functions sensitized to pressure gradient. In *ASME FED. Separated and complex flows, Vol. 217*. ASME.

Launder, B. E., & Spalding, D. B. (1974). The numerical calculations of turbulent flows. *Computational Methods of Applied Mechanical Engineering, 3*, 269.

Logtenberg, S. A., Nijemeisland, M., & Dixon, A. G. (1999). Computational fluid dynamics simulations of fluid flow and heat transfer at the wall—particle contact points in a fixed bed reactor. *Chemical Engineering Science, 54*, 2433.

Lopes, R. J. G., & Quinta-Ferreira, Rosa M. (2009). CFD modelling of multiphase flow distribution in trickle beds. *Chemical Engineering Journal, 147*, 342—355.

Magnico, P. (2003). Hydrodynamic and transport properties of packed bed in small tube-to-sphere diameter ratio: pore scale simulation using an Eulerian and a Lagrangian approach. *Chemical Engineering Science, 58*, 5005.

Maier, R. S., Kroll, D. M., Kutovsky, Y. E., Davis, H. T., & Bernard, R. S. (1998). Simulation of flow through bead packs using the lattice-Boltzmann method. *Physics of Fluids, 10*, 60.

Mantle, M. D., Sederman, A. J., & Gladden, L. F. (2001). Single and two-phase flow in fixed-bed reactors: MRI flow visualization and lattice-Boltzmann simulation. *Chemical Engineering Science, 56*, 523.

Martin, J. J., McCabe, W. L., & Monrad, C. C. (1951). Pressure drop through stacked spheres. *Chemical Engineering Progress, 47*(2), 91.

McHyman, J. (1984). Numerical methods for tracking interfaces. *Physica D: Non-linear Phenomena, 12,* 396−407.

Mehta, D., & Hawley, M. C. (1969). Wall effect in packed column. *Industrial and Engineering Chemistry Process Design and Development, 8,* 280−282.

Mewes, D., Loser, T., & Millies, M. (1999). Modelling of two-phase flow in packings and monoliths. *Chemical Engineering Science, 54,* 4729.

Monaghan, J. J. (1994). Simulating free surfaces with SPH. *Journal of Computational Physics, 110,* 339.

Mueller, G. E. (1991). Prediction of radial porosity distribution in randomly packed fixed beds of uniformly sized spheres in cylindrical containers. *Chemical Engineering Science, 46,* 706.

Natarajan, R., & Acrivos, A. (1993). The instability of the steady flow past spheres and disks. *Journal of Fluid Mechanics, 254,* 323.

Nemec, D., & Levec, J. (2005). Flow through packed bed reactors: 1. Single phase flow. *Chemical Engineering Science, 60,* 6947−6957.

Nijemeisland, M., & Dixon, A. G. (2001). Comparison of CFD simulations to experiment for convective heat transfer in a gas−solid fixed bed. *Chemical Engineering Journal, 82,* 231.

Nijemeisland, M., Dixon, A. G., & Stitt, H. (2004). Catalyst design by CFD for heat transfer and reaction in steam reforming. *Chemical Engineering Science, 59,* 5185−5191.

Ranade, V. V. (1994). Modelling of flow maldistribution in a *fixed bed* reactor using PHOENICS. *The Phoenics Journal of Computational Fluid Dynamics and it's Applications, 7*(3), 59−72.

Ranade, V. V. (1997). Improve reactor via CFD. *Chemical Engineer, 104*(May), 96−102.

Ranade, V. V. (2002). *Computational Flow Modeling for Chemical Reactor Engineering.* London: Academic Press.

Rao, V. G., Ananth, M. S., & Varma, Y. B. G. (1983). Hydrodynamics of two phase cocurrent downflow through packed beds. *AIChE Journal, 29*(3), 467.

Rider, W. J., & Kothe, D. B. (1995). Stretching and tearing interface tracking methods. *AIAA Paper,* 95−117, 171.

Rudman, M. (1997). Volume tracking methods for interfacial flow calculations. *International Journal for Numerical Methods in Fluids, 24,* 671.

Saez, A. E., & Carbonell, R. G. (1985). Hydrodynamic parameters for gas liquid cocurrent flow in packed beds. *AIChE Journal, 31,* 52.

Sangani, A. S., & Acrivos, A. (1982). Slow flow through a periodic array of spheres. *International Journal of Multiphase Flow, 8,* 343.

Sederman, A. J., & Gladden, L. F. (2001). Magnetic resonance visualization of single- and two-phase flow in porous media. *Magnetic Resonance Imaging, 19,* 339.

Sederman, A. J., Johns, M. L., Bramley, A. S., Alexander, P., & Gladden, L. F. (1997). Magnetic resonance imaging of liquid flow and pore structure within packed bed. *Chemical Engineering Science, 52,* 2239.

Seguin, D., Montillet, A., & Comiti, J. (1998). Experimental characterization of flow regimes in various porous media − I: limit of laminar flow regime. *Chemical Engineering Science, 53*(21), 3751.

Seguin, D., Montillet, A., Comiti, J., & Huet, F. (1998). Experimental characterization of flow regimes in various porous media − II: transition to turbulent regime. *Chemical Engineering Science, 53*(22), 3897.

SØrensen, J. P., & Stewart, W. E. (1974). Computation of forced convection in slow flow through ducts and packed beds — III. Heat and mass transfer in a simple cubic array of spheres. *Chemical Engineering Science, 29,* 827.

Specchia, V., & Baldi, G. (1977). Pressure drop and liquid hold up for two phase co-current flow in packed bed. *Chemical Engineering Science, 32,* 515.

Spedding, P. L., & Spencer, R. M. (1995). Simulation of packing density and liquid flow in fixed beds. *Computers and Chemical Engineering, 19,* 43.

Stephenson, J. L., & Stewart, W. E. (1986). Optical measurements of porosity and fluid motion in packed beds. *Chemical Engineering Science, 41,* 2161.

Suekane, T., Yokouchi, Y., & Hirai, S. (2003). Inertial flow structures in a simple-packed bed of spheres. *AIChE Journal, 49,* 1.

Szady, M. J., & Sundaresan, S. (1991). Effect of boundaries on trickle bed hydrodynamics. *AIChE Journal, 37,* 1237.

Thompson, K. E., & Fogler, H. S. (1997). Modeling flow in disordered packed beds from pore-scale fluid mechanics. *AIChE Journal, 43*(6), 1377.

Tobis, J. (2000). Influence of bed geometry in its frictional resistance under turbulent flow conditions. *Chemical Engineering Science, 55,* 5359.

Unverdi, S. O., & Tryggvason, G. (1992). A front tracking method for viscous, incompressible, multi-fluid flows. *Journal of Computational Physics, 100,* 25.

Van Hasselt, B. W., Lindenbergh, D. J., Calis, H. P., Sie, S. T., & Van Den Bleek, C. M. (1997). The three-levels-of-porosity reactor. A novel reactor for countercurrent trickle-flow processes, *Chemical Engineering Science, 52*(21−22), 3901−3907.

Wolfstein., M. (1969). The velocity and temperature distribution of one-dimensional flow with turbulence augmentation and pressure gradient. *International Journal of Heat and Mass Transfer, 12,* 301.

Yin, F. H., Sun, C. G., Afacan, A., Nandakumar, K., & Chung, K. T. (2000). CFD modeling of mass-transfer processes in randomly, packed distillation columns. *Industrial and Engineering Chemistry Research, 39,* 1369.

Zeiser, T., Steven, M., Freund, H., Lammers, P., Brenner, G., Durst, F., et al. (2002). Analysis of the flow field and pressure drop in fixed-bed reactors with the help of lattice Boltzmann simulations. *Philosophical Transactions of the Royal Society of London, Series A: Mathematical, Physical and Engineering Sciences, 360,* 507.

Chapter 5

Reactor Performance and Scale-Up

Nothing in this world is to be feared. It is only to be understood.

Marie Curie

INTRODUCTION

Design, scale-up, and improving performance of a large-scale trickle bed reactor is essentially a multistep and multitask process. Issues, models, and methodologies discussed in Chapters 2–4 need to be used together to achieve this goal. A suggested approach is schematically shown in Fig. 1. Broadly the process involves the following steps:

- Estimate flow regimes, hydrodynamics, maldistribution, mixing (Chapter 2)
- Analyze reaction kinetics, effective rate, mass and heat transfer effects (Chapter 3)
- Establish relationship between hardware and performance (Chapter 4)

Information and correlations discussed in Chapter 2 can be used to identify prevailing flow regimes and estimate key design parameters. Using the basic reaction engineering models discussed in Chapter 3, reactor volume and operating conditions (gas and liquid flow rates, pressure, temperature) can be estimated for the desired conversion and efficiency. Often formulation of these models is based on many assumptions and the design parameters are calculated from the empirical correlations. These correlations are usually obtained from the cold flow experimental data collected with small-scale laboratory reactors using the air–water system. Therefore, extension of these models for realistic cases, scale-up, and designing of actual reactor hardware involves uncertainties. The computational fluid dynamics (CFD)-based models discussed in Chapter 4 can be used to understand important design aspects and reduce the uncertainties. Because of inherent complexities of multiphase flow through packed beds, the computational flow models will have their uncertainties as well. In this chapter, we discuss the overall approach and how models and information

Trickle Bed Reactors. DOI: 10.1016/B978-0-444-52738-7.10003-8

171

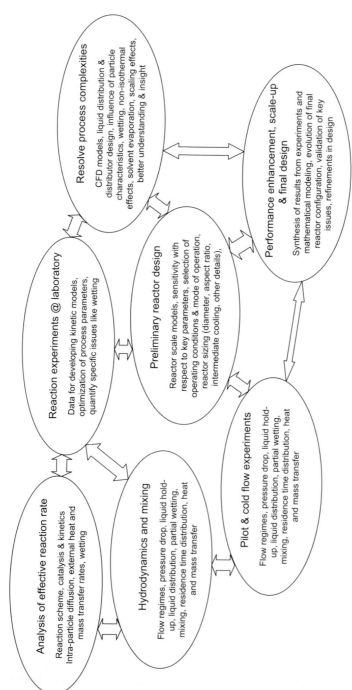

FIGURE 1 Schematic of the overall methodology for relating performance of the reactor with the hardware and the operating protocol.

discussed in Chapters 2–4 can be used to design and scale-up and to enhance performance of trickle bed reactors.

The following section discusses various issues affecting reactor performance. The overall reaction rate is affected by one or more steps (kinetics, mass transfer, heat transfer) and critical analysis of these steps is the key challenge in evaluating the performance of a reactor. The basic rate analysis discussed in Chapter 3 is extended here to include practical issues like influence of particle diameter, operating conditions (flow rate, temperature, and pressure), hysteresis, and so on. Aspects of liquid distribution, mixing, and residence time distribution (RTD) (including a brief summary of different distributors) are discussed in *Reactor Performance*. Key issues relevant to design and scale-up of trickle bed reactors are discussed in *Trickle Bed Reactor Design and Scale-Up*. A strategy for combining the models and methods discussed in Chapters 2–4 with the key aspects of practical trickle bed reactor engineering is discussed in the *Engineering of Trickle Bed Reactors* section. A summary is provided at the end.

REACTOR PERFORMANCE

Effective Reaction Rate and Performance

Effective reaction rate is a function of various parameters such as interphase heat and mass transfer, intraparticle diffusion, and intrinsic reaction kinetics. In some situations, the effective rate is predominantly limited by one of these processes; allowing a simplified analysis of reactor performance. However, more often than not, effective rate is controlled by more than one processes. Interdependence among various factors is shown schematically in Fig. 2. Coupling between mass and heat transfer with wetting of catalyst particles can further complicate the determination of controlling steps and effective rates. It is not uncommon that reaction rates are kinetically controlled in some region of the reactor and mass transfer controlled in the other.

The first step in carrying out effective rate analysis is based on intrinsic kinetics assuming that the reactor is operated under ideal conditions such as absence of any transport limitations, complete wetting of catalyst particles, plug flow operation, and isothermal conditions. This provides a baseline with which influence of other transport-related parameters on effective rate is carried out:

- Gas–liquid mass transfer
- Gas–solid mass transfer over unwetted catalyst particles
- Liquid–solid mass transfer
- Diffusion of reactants inside the particle
- Reactions on the catalyst surface

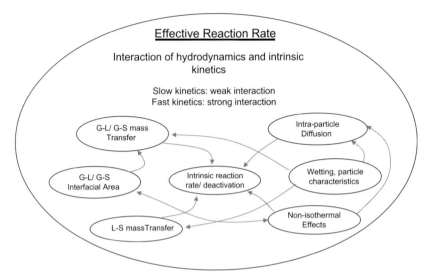

FIGURE 2   Interdependence of hydrodynamic and reaction parameters in trickle bed reactors.

Such studies may provide insights into elementary processes useful for devising design guidelines and operating protocols. In the following sections, influence of various operating parameters on effective reaction rate is discussed.

If *intraparticle diffusion step* is limiting the overall reaction rates, one would get lower conversion in the actual reactor than kinetically controlled conditions. Such a situation occurs for many chemical reactions carried out in trickle bed reactors (Herskowitz & Mosseri, 1983) and the catalyst effectiveness factor was found to be in the range of 0.01–0.3. Decrease in the particle diameter improves the catalyst effectiveness factor considerably. Influence of particle diameter on effective (or global) reaction rate is illustrated in Fig. 3 and as expected, the global reaction rate is significantly higher for smaller particles. However, decrease in particle size may alter the performance of the reactor due to increase in wetting characteristics of the particles. Depending upon whether limiting reactant is in the gas or liquid phase, the performance of the reactor may increase or decrease with reduced wetting.

An increase in superficial *gas velocity* leads to decrease in the liquid holdup and therefore reduces overall conversion (reducing effective residence time for liquid phase). The gas velocity, however, has negligible influence on effective reaction rate (Goto & Smith, 1975; Julcour, Jaganathan, Chaudhari, Wilhelm, & Delmas, 2001; Rajashekharam, Jaganathan, & Chaudhari, 1998) provided that gas–liquid mass transfer is not controlling. In cases where gas–liquid mass transfer rate governs the effective rate, increase in gas

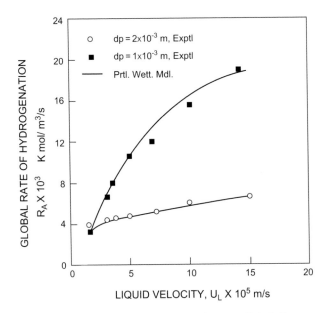

FIGURE 3 Effect of particle diameter on global reaction rate (Rajashekharam et al., 1998). System: hydrogenation of 2,4-dinitrotoluene on 5% Pd/Al$_2$O$_3$; $C_L = 0.2$ kmol/m^3; H$_2$ pressure = 1.4 MPa; $U_G = 4.23 \times 10^{-3}$ m/s; $T = 318$ K.

velocity enhances the reaction rates by improving gas–liquid mass transfer rate as observed by Metaxas and Papayannakos (2006) for hydrogenation of benzene.

The effect of *liquid velocity* on the effective reaction rate is more complex and not monotonic. Increase in liquid superficial velocity leads to shorter residence time and therefore reduces conversion (see, for example, Chaudhari, Jaganathan, Mathew, Julcour, & Delmas, 2002; Khadilkar, Wu, Al-Dahhan, Dudukovic, & Colakyan, 1996; Liu, Mi, Wang, Zhang, & Zhang, 2006; Rajashekharam et al., 1998). However, effective reaction rate may increase with liquid velocity because of enhanced wetting of catalyst particles. An example of this is shown in Fig. 4 for the reaction of hydrogenation of dicy-clopentadiene. However, if the reaction is controlled by gas phase reactant, then the influence of liquid velocity on the rate is opposite, i.e., reaction rate decreases with liquid velocity. In such cases, enhanced wetting due to higher liquid velocity reduces a direct contact of gas phase reactant with the catalyst surface and hence reduces the overall reaction rate (see, for example, cases discussed by Chaudhari et al., 2002; Herskowitz & Mosseri, 1983; Ravindra, Rao, & Rao, 1997; and Fig. 11 in Chapter 3).

Increase in operating pressure or temperature generally enhances the conversion and effective reaction rates. Increase in pressure results in increased concentration of gas phase reactants and higher solubility. Solvent evaporation

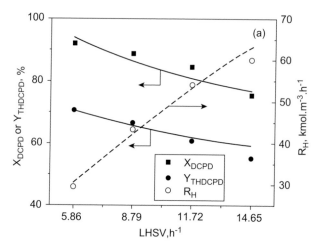

FIGURE 4   Effect of liquid velocity on conversion of dicyclopentadiene (DCPD) and tetrahydrodicyclopentadiene (THDCPD) and reaction rates (Liu et al., 2006). System: hydrogenation of dicyclopentadiene on 0.3%Pd/Al$_2$O$_3$; $T_0 = 348.15$ K; $P = 1.0$ MPa; LHSV = 8.79 h^{-1}.

rate decreases with increase in the operating pressure which may result in higher temperatures within the bed. Increase in operating or prevailing temperature in the bed increases intrinsic reaction rates but with increase in temperature, mass transfer limitations can become significant. The influence of temperature on effective reaction rate can be mixed. If the solvent is very volatile, increase in temperature will lead to increased partial pressure of the solvent in gas phase which may reduce effective reaction rate as well as lowering of temperature rise due to evaporation. The vaporization may lead to partial drying of catalyst surface triggering the gas–solid reactions and multiplicity behavior due to simultaneous effect of exothermicity of the reactions. Such an effect is reported in many studies (Hanika, Lukjanov, Kirillov, & Stanek, 1986; Hanika, Sporka, Ruzicka, & Krausova, 1975; Hanika, Sporka, Ruzicka, & Pistek, 1977; Kheshgi, Reyes, Hu, & Ho, 1992; Kirillov & Koptyug, 2005; Shigarov, Kuzin, & Kirillov, 2002) and is illustrated in Fig. 11 of Chapter 3 for the case of hydrogenation of cyclohexene at different temperatures. In this case, increase in operating temperature increases rate of liquid evaporation as well as dryness of the catalyst surface.

For non-volatile liquid reactant, reduction in wetting surface area results in deterioration of performance of the reactor due to decrease in liquid–solid contacting efficiency. For volatile liquid reactants, increase in dry surface enhances the reactor performance up to a limit because of direct gas phase reactions through dry surface with liquid reactants present in porous catalyst by capillary forces. Effect of increase in temperature on reactor performance for hydrogenation of cyclohexene is shown clearly by results (Fig. 5) of Hanika

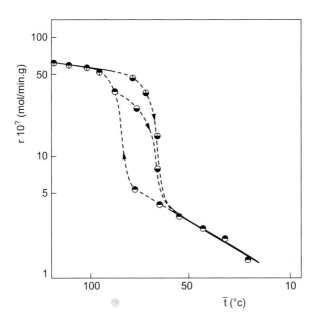

FIGURE 5 Reaction rate variation due to phase transition (Hanika et al., 1975). System: hydrogenation of cyclohexene on 3% Pd/activated C, *Upper curve conditions*: $CL = 5.9$ wt%, $F_L = 1$ cm^3/s, $F_{H_2} = 40$ cm^3/s; *inner lower curve conditions*: $CL = 25$ wt%, $F_L = 4.5$ cm^3/s, $F_{H_2} = 760$ cm^3/s.

et al. (1975). Increase in temperature results in phase change and discontinuity in the variation of rate with temperature since the reactions on gas-covered catalyst surface are significantly faster. In the range of 65–120 °C, phase transition occurs and both liquid phase and gas phase reactions contribute to the effective rate. The evaporation and condensation of liquid phase reactants in this case also result in a distinct hysteresis behavior along the paths of increasing and decreasing operating temperature.

The hysteresis is also observed for hydrogenation of 2,4-dinitrotoluene (see Fig. 6) even if there is no significant evaporation of the liquid phase. This physical phenomenon is attributed to the effect of exothermicity of the reaction and thermal inertia of particles. Along the path of decreasing liquid velocity, the particles are hotter than those along the path of increasing liquid velocity. This leads to higher reaction rates (Fig. 6) along the path of decreasing liquid velocity. Figure 7 illustrates the effect of change in hydrogen pressure on temperature and therefore on the reaction rates. For the liquid-filled catalyst, increase in hydrogen pressure follows lower branch of the temperature where liquid phase reactions are dominant. When hydrogen partial pressure is decreasing, it follows the upper branch of temperature where occurrence of condensation is not sufficient to mimic particle wetting conditions along the path of increasing hydrogen partial pressure. This pore-scale phenomenon is associated with an intricate interaction between heat, mass transfer, and capillary action inside the porous catalyst particle which is demonstrated in several theoretical and experimental studies (Hessari & Bhatia, 1997; Kim & Kim, 1981; Watson & Harold, 1993).

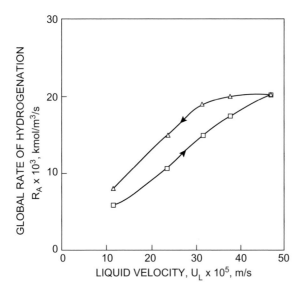

FIGURE 6 Hysteresis of reaction rates associated with hydrodynamics of the bed (Rajashekharam et al., 1998). System: hydrogenation of 2,4-dinitrotoluene; Catalyst: 5% Pd/Al$_2$O$_3$; $C_L = 0.5$ kmol/ m^3, H$_2$ pressure = 3 MPa; $U_G = 4.23 \times 10^{-3}$ m/s; $T = 328$ K.

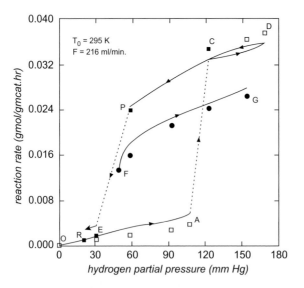

FIGURE 7 Hysteresis of reaction rate with hydrogen partial pressure (Hessari & Bhatia, 1997). System: hydrogenation of cyclohexene; Catalyst: 0.5%Pd/ γ-alumina; $T = 295$ K; $F_L = 216$ ml/m^3.

Flow maldistribution at bed scale, liquid spreading at particle scale, and internal catalyst wetting at pore scale interact with the heat and mass transfer with intrinsic reaction kinetics to influence the effective reaction rates in a complex way. It is difficult to quantify these interactions and their influence on effective rates without a proper mathematical model. Advanced imaging techniques like MRI enable direct visualization of these transport processes occurring at different scales which is illustrated in Fig. 8. Images shown in

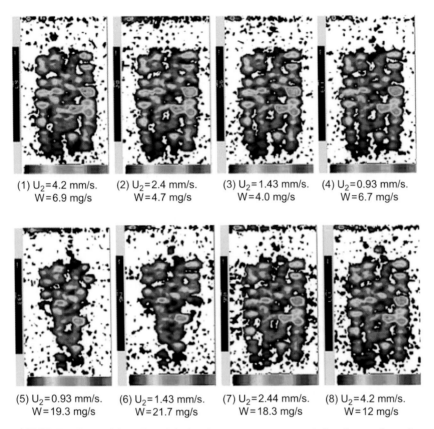

(1) U_2=4.2 mm/s.   (2) U_2=2.4 mm/s.   (3) U_2=1.43 mm/s.   (4) U_2=0.93 mm/s.
   W=6.9 mg/s   W=4.7 mg/s   W=4.0 mg/s   W=6.7 mg/s

(5) U_2=0.93 mm/s.   (6) U_2=1.43 mm/s. (7) U_2=2.44 mm/s.   (8) U_2=4.2 mm/s.
   W=19.3 mg/s   W=21.7 mg/s   W=18.3 mg/s   W=12 mg/s

FIGURE 8 The particle-scale and bed-scale transport processes during the reaction using MRI techniques in the form of liquid distribution contours (Kirillov & Koptyug, 2005, U_2 = liquid superficial velocity, W is apparent output of trickle bed). System: hydrogenation of α-methyl styrene to cumene and heptene on 1%Pd/γ-Al$_2$O$_3$, +1%Mn with hydrogen velocity 10–50 cm/s.

Fig. 8 represent the fraction of liquid present inside the trickle bed reactor at various liquid flow rates. Images 1–4 indicate the variation of the liquid fraction with decreasing liquid flow rate. Fraction of the wetted particles decreases with the decreasing liquid flow rate. Further decrease in the flow rate below 0.93 mm/s leads to dry out phenomenon and runaway conditions. Images 5–8 indicate variation of the liquid fraction with increasing liquid flow rate. Increase in liquid flow rates at this point leaves the central bed in wetted condition while leaving rest of the bed in dry condition. Reactor performance was found to be the highest at this point due to the reactions on gas-covered surface of the catalyst (Kirillov & Koptyug, 2005). Comparison of the images at the same liquid velocities suggests that hysteresis behavior is present at all different scales of the reactor and development of suitable modeling techniques to describe these phenomena is necessary to avoid unpleasant surprises.

Particle Characteristics

Key characteristics of trickle bed reactors depend on packing of particles, which in turn depends on particle characteristics. Apart from selection of the catalyst, design of catalyst particle (shape and size) suitable for reactor operation is crucial for efficient and successful operation of the reactor. The key objective of designing catalyst particle size and shape is to accommodate maximum amount of catalyst per unit volume of the reactor without adversely affecting other factors such as pressure drop, liquid holdup, interfacial area, wetting, overall effectiveness factor of the catalyst, thermal, and mechanical aspects of the catalyst. Hydrodynamic parameters are function of underlying fluid dynamics in the bed void. How particle size, shape, and their arrangements influence the local fluid flow is discussed in Chapter 4 (Characterization of Packed Beds). Apart from fluid dynamics of the packed bed, particle size and shape significantly affect catalyst effectiveness factors and thereby effective reaction rates. This section discusses some of the key issues in selecting catalyst particle size and shape and their influence on performance of trickle bed reactors.

Particle characteristics are determined in terms of the following:

- Particle size (diameter, equivalent diameter, volume equivalent diameter, and size distribution)
- Shape of the particle (measured in terms of sphericity)
- External surface area (surface area per unit volume of the catalyst)
- Internal pore diameter (and pore size distribution)
- Internal surface area (internal surface area per unit volume of catalyst particle)
- Other parameters: density, surface roughness, wettability, and hardness/crushing strength

Bed characteristics affected by selection of the particles are as follows:

- Porosity (volume of catalyst particles per unit volume of bed) and tortuosity of the bed
- Specific surface area of reactor (external catalyst surface area of bed per unit volume of reactor)
- Bulk density of bed (total weight of catalyst per unit volume of reactor)

Conventionally spherical, cylindrical, and ring type of catalyst particles were obvious choices mainly due to easy manufacturability. In recent years, various other shapes such as trilobe, hollow cylinders, quadrilobes, and poly-lobes have been developed to enhance interface area, catalyst effectiveness, and mechanical strength (Sie, 1993). For the same outer diameter, trilobes or wagon wheels may provide more than twice surface area compared to the conventional spherical or cylindrical particles. For trickle bed operations, trilobes and quadrilobes are preferred (because of larger catalyst loading and less likelihood of stagnant regions) over hollow cylinder particles for avoiding stagnant region and larger catalyst loading. Bruijn, Naka, and Sonnemans (1981) have demonstrated that activity (effectiveness) of polylobes is higher than cylindrical particle for equal volume of catalyst particle. From catalyst effectiveness point of view, smaller particles are preferred over larger particles. However, permissible pressure drop across the bed defines a lower limit on how smaller-sized catalyst particles can be used.

Influence of particle shape and size on overall bed characteristics (voidage and interfacial surface area) is reported in Table 1. Apart from the size and shape, the method of packing catalyst particles in a bed may have a significant impact on performance. Among various packing methods

TABLE 1 Typical Characteristics of Particles Used in Trickle Bed Reactor

Shape	Size (mm)	Specific Area (m^2/m^3)	Bed Porosity	Source
Spherical	2		0.38	Larachi et al. (1991)
	3	1200	0.40	
	6	620	0.38	
	2.4		0.39	
	1.4		0.35	Larachi et al. (1991)
Cylindrical	$1.6d_p \times 4.7l_p$		0.437	Nemec and Levec (2005)
	$3.2d_p \times 3.4l_p$	1083	–	
Rasching rings	$10 \times (6.5 \times 5)$	924	0.67	
Extrudate	$1.5d_p \times 3.11l_p$		0.40	
	$3.17d_p \times 9.5l_p$		0.37	
	$1.2d_p \times 3l_p$		0.37	
Trilobes	$1.27d_p \times 5.5l_p$ ($d_e = 1.41$)		0.466–0.511	Nemec and Levec (2005)
	1.27 ($d_{eqv} = 3.4$ mm)	235 m^2/g	0.53	
Quadrilobes	1.35×5.2		0.593	Nemec and Levec (2005)

reported by Al-Dahhan, Wu, and Dudukovic (1995), random loose packing (RLP) and random close packing (RCP) are the two commonly used methods for packing catalyst bed. Porosity variation due to these two packing methods can be as large as 16–20% (Klerk, 2003). RCP can be achieved by tapping bed continuously while filling the bed with catalyst particles. Trickle bed hydrodynamics and performance is quite sensitive to the bed porosity and therefore it is important to quantify bed porosity accurately while characterizing the catalyst bed. Particle density, surface roughness, wettability, and hardness/crushing strength are also important characteristics which influence hydrodynamic interactions. Surface properties of particles can be manipulated to achieve better catalyst surface wetting and better spreading of liquid. Effective wetting of catalyst bed also depends on liquid distribution. Some of the distributor designs used in practice are briefly discussed in the following.

Gas–Liquid Distributor

Liquid distribution in trickle bed reactors significantly influences its performance especially for liquid reactant-limiting reactions. Poor liquid distribution can lead to large gas or liquid pockets resulting in a poor overall external mass transfer of gas or liquid reactants to the catalyst surface and hence to lower reactor performance. If reaction is exothermic and comprises volatile liquid components, the gas phase reactions in non-wetted region cause formation of local hotspots leading to catalyst deactivation. Therefore, uniform distribution of the liquid at the inlet as well as in the bed is essential for achieving better performance of the reactor. Proper distribution at the inlet of the bed is one of the effective ways to minimize adverse effects of the liquid maldistribution. Despite uniform distribution at the inlet, liquid maldistribution may occur along the length of the column because of other factors such as packing characteristics and bed tilt. In such cases redistribution of liquid after certain height of the bed is necessary to control the liquid maldistribution. It is important to note that the porous bed of catalyst particles itself facilitates liquid distribution to a certain extent but if a proper distributor at the inlet is not used, a significant portion of the bed near the liquid inlet may remain unwetted.

Desired characteristics of liquid distributor are as follows:

- low pressure drop
- ability to operate over a broad range of gas–liquid flow rates
- less susceptibility to blockage and fouling
- ease of operation
- ease of installation and maintenance (cleaning)

Different types of liquid distributors used in trickle bed reactor are shown in Fig. 9. In many fine chemical applications, gas and liquid flow rates are

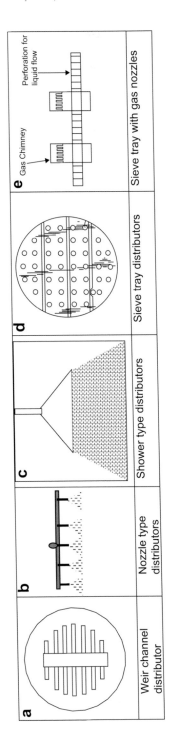

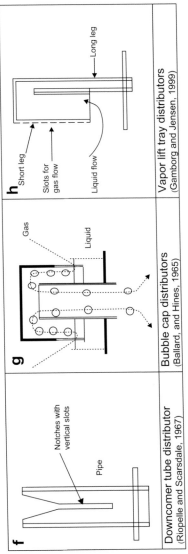

FIGURE 9 Different types of liquid distributors used in trickle bed reactors.

moderate and the diameter of the column is rather small. For such reactors, simple distributor design such as sieve tray or layer of inert particles on the upper portion of the bed may serve the purpose. Resistance offered by the sieve tray or top layer of particles is adequate to spread the liquid radially across the bed cross-section. In petroleum and petrochemical applications (like hydro-processing or hydrocracking processes), reactor diameters are large (~ 6 m). Relatively, lower liquid flow rates further lead to difficulties in achieving uniform liquid distribution. Specially designed distributors are needed to achieve the uniform liquid distribution in such large column diameter reactors. In such cases, even a small tilt in the reactor orientation can lead to severe liquid maldistribution.

Weir channel-type distributors (as shown in Fig. 9a) are commonly used in many low to moderate diameter trickle bed reactors. In this type, liquid is centrally distributed in several radially arranged weirs which have triangular notches. These distributors offer very low pressure drop and the gas flows independently without much interaction with the liquid phase. Liquid distri-bution is affected by a number of weirs and number of notches provided in such distributors. The degree of distribution is sensitive to the correct orientation (deviation from horizontal level) of weir channels. Wide range of liquid flow rates can be handled with these distributors.

Nozzle-type and shower-type liquid distributors (see Fig. 9b and c) are generally used in small diameter columns due to large pressure drop associated with these columns. These distributors are not sensitive to the bed tilt. For smaller diameter columns, spray-type distributor is used and for medium diameter columns nozzle-type distributors are used. These distributors are not suitable for foaming liquids and low liquid flow rates.

Sieve (perforated) tray and sieve tray with gas nozzles (refer Fig. 9d and e) are the most commonly used distributors in trickle bed reactors. If gas flow rate is low to moderate, sieve tray distributor is used. For higher gas flow rates, sieve tray with gas nozzle is generally used to reduce pressure drop. These distrib-utors are suitable for low as well as high gas–liquid flow rates. For high liquid flow rates, liquid flow can occur through weirs/notches of the gas nozzles. These distributors are also suitable for foaming liquids and are sensitive to the orientation of the tray.

The downcomer-type tube distributors (refer Fig. 9f) are suitable for large diameter columns and foaming liquids. In these distributors, liquid height is maintained on the perforated trays. Gas passes through the center of the vertical tube and liquid passes through the vertical notch on the wall of the tube. Liquid flow rate is proportional to the height of liquid pool. These distributors are moderately sensitive to the level of the tray.

In bubble cap distributors (shown in Fig. 9g), gas bubbles pass through the liquid pool maintained by a riser tube. These bubbles carry liquid along with it which flows from bottom of the tray to the vertical notches on riser tube. Through downcomer tube, vapor liquid mixture disengages in the bed. Bubble cap distributor thus provides intimate gas–liquid contacting. This type of distributor can handle a wide range of gas and liquid flow rates and is suitable for large diameter columns. Pressure drop and sensitivity to the tilting of the tray are also moderate for this distributor.

Construction of vapor lift tube distributor (as shown in Fig. 9h) is simpler than the bubble cap tray distributor and requires less space. This may allow use of more number of tubes to improve uniformity of the liquid distribution. Here, liquid height is maintained above the side cap so that vapor pressure outside the cap raises the liquid above riser tube with intimate contact with the gas. This type of distributor offers similar advantages to that of bubble cap distributor.

Various factors such as dimension of bed, gas–liquid flow rates, properties of the fluid (viscous, foaming, fouling, etc.), and pressure drop need to be considered while selecting the type of distributor. Influence of liquid distribution on the reactor performance is discussed in the next section.

Liquid Maldistribution and Performance

Trickle bed reactors are generally operated at low liquid velocities; therefore inlet distributor and local bed properties play an important role in liquid distribution inside the bed. Liquid maldistribution directly affects the performance due to improper contacting of gas–liquid phases over catalyst surface and channeling of the flow. Gravity-driven nature of liquid flow offers relatively few degrees of freedom to tune/manipulate the liquid distribution. Liquid maldistribution in the trickle bed reactors can be classified into two categories: gross maldistribution and local maldistribution. Improper liquid distribution at the inlet causes gross maldistribution which can be minimized by proper design of the distributor (discussed in the previous sub-section). On the other hand, local maldistribution may occur due to various factors such as properties of particles (size, shape, surface roughness, etc.), arrangement of particles, packing density, and properties of the gas and liquid phases. In this section, details of local liquid maldistribution and its implications on reactor performance are discussed.

In many studies on trickle beds, influence of liquid distribution was accounted in terms of wetting efficiency (see, for example, Herskowitz & Smith, 1978; Sylvester & Pitayagulsarn, 1975; Weekman & Myers, 1964). However, subsequent experiments as well as mathematical modeling indicate that the liquid distribution effects are more significant than the wetting effects (modeling: Crine & L'Homme, 1984; Crine, M., Marchot, P., & L'Homme 1981; Fox, 1987; Funk, Harold, & Ng, 1990; experimental: Borda, Gabitto, & Lemcoff, 1987; Chou, 1984; Lutran, Ng, and Delikat, 1991). Quantification of maldistribution within the bed is rather difficult. Usually liquid distribution at the outlet is used to quantify liquid maldistribution (Borda et al., 1987; Herskowitz & Smith, 1978). However, usefulness of the above method is restricted to smaller-sized reactors. Fundamental understanding of liquid distribution and spreading within packed

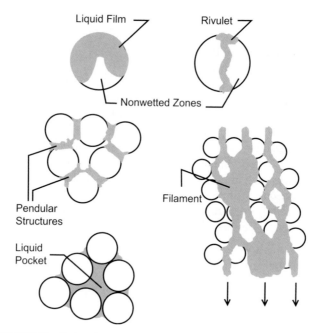

FIGURE 10   Various liquid flow patterns proposed by Lutran et al. (1991).

beds gained importance in recent years. Using computer-assisted tomographic (CT) technique, Lutran, Ng, and Delikat (1991) have demonstrated various interesting flow features of the trickle flow regime (see Fig. 10).

Depending upon the flow rates and properties of the fluid and particles, liquid in the void may be present in the form of films, rivulets, pendular structure, filament, or stagnant pockets. Film flow and rivulets phenomena are associated with a single particle; while in multiparticle systems, various patterns can be observed such as pendular structure, filament, and stagnant liquid pockets. Pendular structure and stagnant liquid pockets are associated with the contact point between the particles and the capillary pressure. Filaments maintain the continuous liquid flow over the particle which can be in the form of liquid film or the rivulets. Formation of the filament-type flow is more likely when bed is not prewetted. The film flow is expected when bed is prewetted (see experimental results shown in Fig. 11).

Rivulet-type flow was also observed by Sederman and Gladden (2001) using MRI scanning method for non-prewetted bed. In non-prewetted bed, liquid preferentially flows near the wall. A small fraction of the flow occurs at the center in the form of filaments. Liquid filaments grow in size with increase in the liquid flow rates. This type of flow produces higher liquid velocity channels flowing in the confined region, where very small fraction of the column is wetted by the liquid. However, in the prewetted bed, spreading of the liquid enhances the wetting fraction of the bed considerably. Implications of

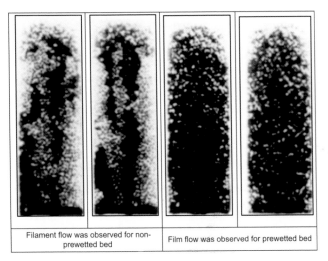

| Filament flow was observed for non-prewetted bed | Film flow was observed for prewetted bed |

FIGURE 11 Liquid distribution in trickle bed characterized using CT (Lutran et al., 1991). System: 3 mm of glass sphere as the particles and water as fluid with uniform distributor.

the local liquid flow structures are also reflected in the global hydrodynamic parameters. For example, filament flow has a little interaction with the gas as compared with the film flow which results in lower pressure drop. Scanning of the flow profile inside the bed provides valuable insight into the complex nature of the flow. In the industrial trickle bed reactors, combination of these flow patterns and their quantification is a major challenge.

Quantitative description of non-uniform liquid distribution can be performed by using the liquid collectors at the outlet of the bed. The use of concentric cylinders at the outlet to measure radial distribution of liquid was found to be a suitable method (Borda et al., 1987; Herskowitz & Smith, 1978; Marcandelli, Lamine, Bernard, & Wild, 2000). Their analysis suggests that radial liquid distribution changes along with the height of the column. For small diameter columns, radial spreading occurs earlier than the large diameter columns. Cylindrical particles produce excessive wall flow compared to the spherical particles. Further detailed information on liquid non-uniformities can be gained using azimuthally divided concentric cylinders (Marcandelli et al., 2000; Møller, Halken, Hansen, & Bartholdy, 1996). Using such approach, Marcandelli et al. (2000) have defined maldistribution factor as follows:

$$M_f = \sqrt{\frac{1}{N(N-1)} \sum \left(\frac{Q_{Li} - Q_{mean}}{Q_{mean}}\right)^2} \qquad (1)$$

where, M_f is the maldistribution factor, N is the number of channels used for collecting liquid over the bed cross-section, and Q_{Li} and Q_{mean} are the flow rate of liquid in each channel and mean flow rate, respectively. A sample of

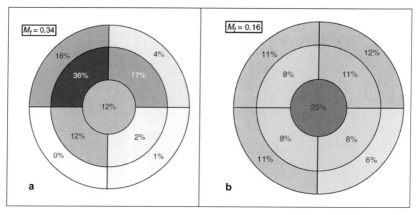

FIGURE 12   Liquid distribution at the outlet without gas flow rate ($U_L = 8$ mm/s) and with gas flow rate = 5 cm/s ($U_L = 3$ mm/s), for column diameter = 0.1 m (Marcandelli et al., 2000).

their results for different gas–liquid velocities is shown in Fig. 12. The studies of Borda et al. (1984) and Sylvester and Pitayagulsarn (1975) suggest that gas flow has a significant effect on liquid distribution inside the bed and increase in gas velocity leads to increase in local liquid flow through the central region. The observations of Møller et al. (1996) also support this finding.

Modeling of liquid distribution in packed bed is rather difficult task due to complexity of packing structure and its interaction with the flowing fluid. Herskovitz and Smith (1978) have developed interconnecting cell model for prediction of liquid distribution. The proposed model predicts the key characteristics of liquid flow and demonstrates the effect of the particle and bed diameter and liquid phase properties. Funk et al. (1990) have modeled two-dimensional packed bed including the distributor and particles. Their model predicts the liquid flow distribution with single and multiple liquid inlets. However, a detailed representation of the fluid–fluid and fluid–solid interactions is lacking in these approaches.

Recently, models based on CFD were proposed for understanding of the liquid distribution in trickle bed reactors (see, for example, Jiang, Khadilkar, Al-Dahhan, & Dudukovic, 1999, 2001; Gunjal, Ranade, & Chaudhari, 2003; Sun, Yin, Afacan, Nandakumar, & Chuang, 2000). Liquid distribution for single, two, and uniform inlets were predicted with fairly good accuracy using discrete cell modeling approach (Jiang et al., 1999, 2001). This model captures key features like formation of liquid channels at point sources and meandering and merging of channels within the bed. For a non-uniform inlet, complete three-dimensional modeling approaches are essential. Such approach was used by Gunjal et al. (2003) for simulating influence of uniform and non-uniform inlets on liquid distribution. A sample of their results is shown in Fig. 13. Their model predicts key features of liquid distribution

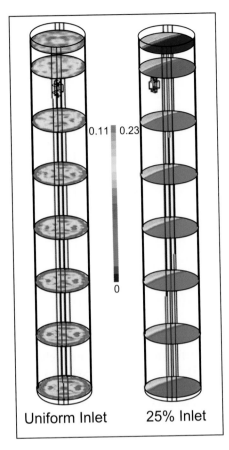

Uniform Inlet   25% Inlet

FIGURE 13 Simulated liquid distribution using three-dimensional CFD model: contours of liquid volume fraction (Gunjal et al., 2003; $D = 0.114$ m and $d_p = 3$ mm; $V_L = 2$ kg/m^{2}s; $V_G = 0.22$ m/s).

observed experimentally. Simulated results also indicate that there is little possibility of achieving better liquid distribution if inlet distribution is not adequate. Such an observation is also reported by the study of Funk et al. (1990) and Marcandelli et al. (2000).

Improper liquid distribution has a direct effect on the performance of the reactor. In maldistributed bed, extent of the bed wetting is poor due to the liquid segregation. Flow is dominated by filament type and merging of these filaments produces channeling effect. Poor wetting leads to under-utilization of catalyst bed. Influence of liquid maldistribution on reactor performance is different for gas-limited reactions and liquid-limited reactions. For liquid-limited reactions, extent of the wetting is directly related to the liquid distribution. Better liquid distribution leads to better liquid–solid contacting and therefore better reactor performance. Better wetting, however, does not necessarily increase effective reaction rate for the cases which are limited by gas phase reactants. This was clearly demonstrated by McManus, Funk, Harold, and Ng (1993) by carrying out the experiments of hydrogenation of

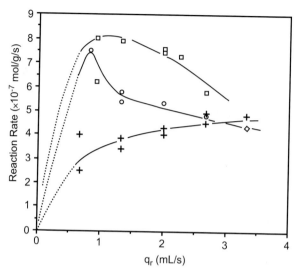

FIGURE 14 Dependence of hydrogenation reaction rate of α-methyl styrene on type of distributor (one tube (+), two tube (○) and four tubes (□); McManus et al., 1993).

α-methyl styrene with single inlet and multiple inlets. This reaction is hydrogen limited. It was observed that effective reaction rates are highest at intermediate wetting condition (see Fig. 14). This is also indirectly observed by Kireenkov, Shigarov, Kuzin, Bocharov, and Kirillov (2006). A sample of their results (for gas phase-limited reaction) which shows influence of column diameter on the reactor performance is shown in Fig. 15 (distribution in larger diameter column may be poorer than the smaller diameter column resulting in better performance of larger reactor).

Recent advances in computational modeling and non-intrusive experimental techniques enable better insights into the liquid distribution in trickle bed reactors. Gross-scale maldistribution is attributed to distributor design while small scale maldistribution is attributed to the local properties of the bed. Though, these sophisticated techniques provide detailed information about flow structures (rivulets, films filaments, stagnancy, etc.) inside the bed, its quantification and use for reaction engineering analysis is still at a preliminary stage.

Residence Time Distribution

RTD has been used for many years to analyze and characterize mixing inside the reactor because of the unavailability of the detailed flow information inside the reactor. Most of the trickle bed reactors are operated at very low liquid flow rates and various studies have shown that dispersion effects are dominant in such cases. Dispersion coefficient lumps influence of complex

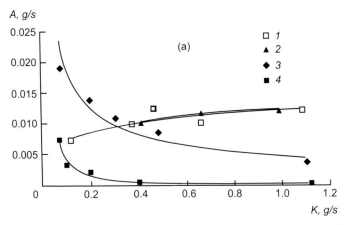

FIGURE 15 Reactor output at different column diameters: (1) 4.2 cm, (2) 2.3 cm, (3) 1.2 cm, and (4) 0.6 cm (Kireenkov et al., 2006). System: hydrogenation of α-methyl styrene on 15.2% Pt/γ-alumina; $T_{in} = 75\ ^\circ$C; $L = 10$ mm; $U_{H_2} = 31.8$ cm/s; $d_p = 1$ mm.

flow non-idealities into a single parameter. Therefore, a careful analysis of RTD data is essential for extracting meaningful information. The work of Van Swaaij, Charpenter, and Villermaux (1969) indicates that the dispersion effects are dominant at lower liquid velocities and follow a relation $Pe \propto Re_L^{\alpha}$, where α is ~ 0.5. Several other studies (Cassanello, Martinez, & Cukierman, 1992; Hochman & Effron, 1969; Saroha, Nigam, Saxena, & Kapoor, 1998; Stegeman, van Rooijen, Kamperman, Weijer, & Westerterp, 1996) have also reported that the gas flow rate has negligible effect on dispersion while particle diameter has a strong influence on dispersion. Therefore, it is convenient to define the dispersion coefficient in terms of particle diameter $Pe_d = Pe d_p/L$. Mears (1971) has proposed a criterion, which relates the minimum length of the catalyst bed beyond which the axial dispersion effects are negligible, as:

$$\frac{L}{d_p} > \frac{20m}{Pe_d} \ln \frac{C_{in}}{C_{out}} \tag{2}$$

where m is the order of the reaction.

Analysis of experimental data by Montagna and Shah (1975) suggests that liquid dispersion effects in residue hydrodesulfurization are equally important as the wetting and liquid holdup; and the Mears criterion was found to be valid for this case.

Thus, RTD is found to be useful for understanding many aspects of trickle bed reactors such as kinetic processes (Ramachandran & Smith, 1978), wetting characteristics (Schwartz, Weger, & Dudukovic, 1976), channeling (Oliveros & Smith, 1982), flow maldistribution (Hanratty & Dudukovic, 1992), and liquid mixing (Gulijk, 1998; Higler, Krishna, Ellenberger, & Taylor, 1999).

Hydrodynamics and mixing in the trickle bed reactors are governed by several factors including non-uniform porosity distribution of the bed, capillary forces, wetting, and non-uniform distribution of flow at the inlet. In Chapter 2, we have seen that prewetted or non-prewetted bed conditions affect the hydrodynamic parameters such as, pressure drop, liquid holdup, and liquid distribution considerably. The implications of these changed hydrodynamic parameters on the mixing of the liquid phase reactants or products can be understood using RTD. RTD or exit age distribution $E(t)$ for the trickle bed reactors under pre-wetted and non-prewetted bed conditions is shown in Fig. 22 of Chapter 2 for 0.1 m diameter column with 3 mm glass particles.

Dispersion is higher for a prewetted bed than non-prewetted bed at lower liquid flow rates. At low liquid flow rate, in the non-prewetted bed, liquid flows through the confined region within the bed due to the capillary effect and therefore, exit age distribution shows dominant plug flow-like characteristics. As liquid flow rate increases, the observed difference in the exit age distributions for the prewetted and non-prewetted conditions reduces. At high liquid flow rates ($10 \, kg/m^2s$), where capillary pressure effect is negligible, RTD is almost independent of extent of prewetting of the bed.

Extent of hysteresis in pressure drop and liquid holdup reduces with increase in particle diameter (Gunjal, Kashid, Ranade, & Chaudhari, 2005). These results indicate that capillary effects are smaller for larger particles. These effects are also reflected in exit age distribution (Gunjal et al., 2003). For a 6-mm particle bed, at low liquid flow rates, liquid dispersion was found to be lower than the 3-mm particle bed. The difference between the exit age distributions for the prewetted and non-prewetted was less for a 6-mm particle bed.

RTD can be simulated by CFD models which can account for details of reactor hardware, porosity distribution, and fluid–fluid interaction. These models can therefore be used to understand influence of various design and operating parameters on RTD. An example of CFD simulations of RTD is shown in Fig. 36 of Chapter 4 in which the model was shown to capture the influence of prewetting on RTD (Gunjal et al. 2003).

Periodic Operation and Performance

The idea of periodic operation of trickle bed reactor was originally derived from the pulse flow operation where gas and liquid phases flow in the high interaction regime. However, realization of pulse flow regime is difficult in the large commercial-scale reactors. Therefore, the concept of forced pulsing or periodic operation (flow modulation) is gaining interest. Periodic operation enhances the gas phase mass transfer due to the direct access of the gaseous reactant to the solid surface in the liquid lean region. This was first demonstrated by Haure, Hudgins, and Silveston (1989) for SO_2 oxidation reaction.

There are many advantages of gas and liquid phase-enriched flow regions which occur in periodic operations. Periodic operation results in higher gas—liquid or liquid mass transfer rates than steady-state operation of the trickle bed reactors. Presence of liquid-enriched zones enhances effective wetting and therefore more effective utilization of the bed. Periodic flow also enhances radial mixing and therefore reduces the extent of liquid maldistribution. Most of the reactions carried out in trickle bed reactors are exothermic. Hence efficient heat removal is essential for effective operation of the reactor. Possibility of liquid maldistribution with the steady-state operation of the trickle bed reactor may lead to existence of hotspots within the bed. Such hotspots influence the quality of the product and may some times cause runaway conditions. In a periodic operation, liquid-enriched zone replenishes the catalyst surface with fresh reactants and therefore significantly reduces problems associated with inadequate wetting, liquid distribution, stagnant zones, and the temperature non-uniformity. Some work on how and when periodic operation can be fruitful is briefly reviewed in the following (for more details see Boelhouwer, Piepers, & Drinkenburg, 2002; Silveston & Hanika, 2002).

Possible benefits of periodic flow operation depend on specific reaction system under consideration. If the reaction is kinetically controlled, periodic flow operation enhances the performance of the reactor compared to the steady-state operation because of better liquid distribution. However, extent of such improvement is rather small because most of the periodic flow operation requires considerably high gas and liquid flow rates. At such high gas and liquid flow rates, the contacting efficiencies are high even for steady operation. Periodic operation is beneficial for *gas-limiting reactions*. For such cases, in steady flow operation, partial wetting conditions are favorable due to direct access of gaseous reactant to the solid surface. However, ineffective heat removal from dry surfaces leads to overheating of catalyst. In contrast to this, periodic operation offers gas-rich zones which enhance the mass transfer effects and liquid-rich zones which effectively remove generated heat. Periodic operation also increases radial mixing and liquid-rich zone replenishes stagnant liquid pockets in the catalyst bed. There are many examples in published literature which demonstrate enhanced performance of trickle bed reactors with periodic operation (see, for example, Castellari & Haure, 1995; Gabarain, Castellari, Cechini, Tobolski, & Haure, 1997; Haure et al., 1989; Lange, Hanika, Stradiotto, Hudgins, & Silveston, 1994; Lee, Hudgins, & Silveston, 1995). A sample of results reported by Khadilkar, Al-Dahhan, and Dudukovic (1999) is shown in Fig. 16.

For *liquid-limiting reactions*, one may not expect significant benefits of the periodic operation over the steady operation. Some results reported by Khadilkar et al. (1999) are shown in Fig. 17 to illustrate this. In fact the periodic operation under-performs the steady-state operation at higher liquid velocities. However, these experiments were carried out using small diameter column

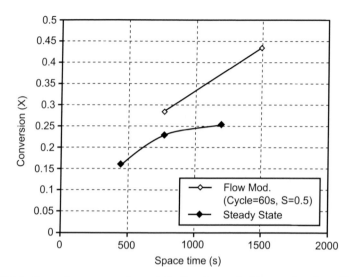

FIGURE 16   Rate enhancement under periodic flow for gas-limiting reaction (from Khadilkar et al., 1999). System: hydrogenation of α-methyl styrene on 0.5%Pd/alumina; $d_p = 3.1$ mm spherical; $T = 20-23$ °C.

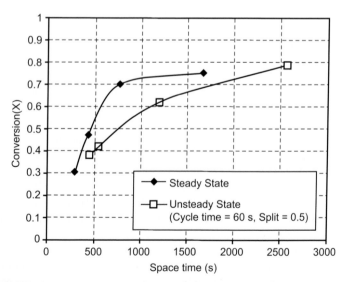

FIGURE 17   Rate enhancement under periodic flow for liquid-limiting reaction (Khadilkar et al., 1999). System: hydrogenation of α-methyl styrene on 0.5%Pd/alumina; $d_p = 3.1$ mm spherical; $T = 20-23$ °C.

(2.2 cm), where liquid distribution and wetting do not affect reactor performance significantly.

TRICKLE BED REACTOR DESIGN AND SCALE-UP

Reactor design involves evolution of reactor hardware and operating protocols which satisfy various process demands without compromising safety, environment, and economics. This therefore necessarily involves expertise from various fields ranging from thermodynamics, chemistry, and catalysis to reaction engineering, fluid dynamics, mixing, and heat and mass transfer. Reactor engineer needs to combine the basic understanding of the chemistry and catalysis with the methods described in this book to evolve suitable reactor design. More often than not, specific scale-up/scale-down methodologies need to be used to establish confidence in the developed design. In this section, some aspects of scale-up are discussed. The overall steps to evolve suitable trickle bed design are discussed in the next section *(Engineering of Trickle Bed Reactors)*.

Reactor Scale-Up/Scale-Down

Scale-up or scale-down is one of the important steps in reactor engineering of trickle bed reactors. Scaling of reactors involves transforming the information from one scale of reactor to another scale (either from a laboratory scale to pilot and commercial-scale reactor or from commercial-scale reactor to pilot and laboratory-scale reactor). Objectives of the scaling may vary from case to case. Scale-up is needed for transforming the newly developed technology at the laboratory-scale reactor to the commercial-scale reactor or simply to interpret and extrapolate laboratory results to the commercial-scale reactor. Reverse case, that is 'scale-down' is also equally important that mimics the large-scale operating process at a much smaller scale. This is used for diagnosis of existing operating reactor using appropriate experiments on the laboratory scale. It may be necessary to use intermediate scales between laboratory and commercial scales especially when transforming information from one scale to the other is not straightforward. Appropriate scaling methodology is required depending upon the specific objectives of the scaling exercise. These objectives could be (i) modification in the existing plant or operational reactor; (ii) increase in the capacity of the existing unit; (iii) modification in the feed or catalyst; and (iv) improvement in the product quality. Overall scaling methodology is schematically shown in Fig. 18.

Scale-up and scale-down methodology needs to estimate how key design parameters change with the reactor scales and then estimate influence of these changes on reactor performance. If the design parameters exhibit simple relationship with the scale of operation of the reactor, then scaling of reactor becomes relatively straightforward. However, for most of the practical

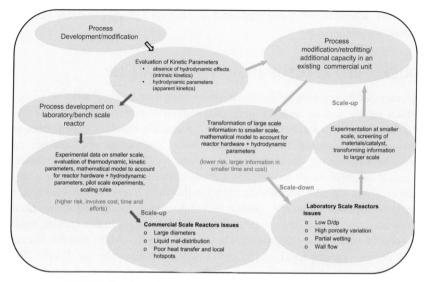

FIGURE 18 General methodology for trickle bed reactor scale-up.

problems this is not the case. Reaction rates, liquid phase mixing, wetting, and heat and mass transfer rates get affected by reactor scales with different degrees and therefore scale-up becomes rather complex. Reaction rate data collected on the laboratory scale forms the basis for scale-up of any process (Fig. 18). Usually the intrinsic reaction rates are studied in a stirred slurry or basket-type reactor. In these reactors, operating conditions such as pressure, temperature, stirring speed, and particle size are chosen such that obtained kinetic data are not influenced by any transport (heat and mass transfer) steps. The apparent reaction kinetics (in the presence of external transport limitations) can be obtained by conducting experiments in the laboratory-scale trickle bed reactor. Sensitivity of reactor performance with operating conditions (such as pressure, temperature, and feed composition) and apparent kinetics can be obtained from such test experiments. These data are useful for optimizing the process conditions to enhance reactor performance. Specific experiments are needed to identify the rate-limiting steps and to quantify influence of operating conditions on wetting, gas—liquid and liquid—solid mass transfer, back mixing, heat transfer, etc. Methods discussed in Chapters 2 and 3 can be used to interpret the laboratory-scale reactor experiments for optimizing operating process conditions. This is a pre-requisite for successful scale-up.

Laboratory-scale reactor can be operated at isothermal conditions which may not be possible for large industrial scale reactors. In small reactors, with lower reactor diameter to particle diameter ratio, liquid bypassing near the wall can affect reactor performance severely. On the other hand, large-scale reactor issues such as liquid segregations, poor heat transfer rates, formation of local hotspots, and rivulet flow are considerably different. There are no general rules which can be applied for scaling-up. The laboratory data and associated mathematical models need to be used to reduce uncertainties associated with scale-up. The overall strategy of engineering of trickle bed reactors is discussed in *Engineering of Trickle Bed Reactors* while the scale-up methodologies are briefly discussed in the following.

Reactor Scale-Up Methodologies

Conventionally scale-up is based on maintaining similarity (with the help of key dimensionless numbers) for different scales of reactors. Similarity could be geometric, kinematic, or thermal. Various stages involved in scale-up process are shown schematically in Fig. 19. For trickle bed reactors, maintaining geometric similarity is quite difficult after certain scale-up ratio (~ 10−20). For high scale-up ratio, reactor becomes too tall which may not be practically feasible. For example, typical hydrotreating reactors may become 500 m tall if geometrical similarity is maintained with the laboratory-scale reactors.

Kinematic similarity involves maintaining the same dimensionless numbers based on velocities in the reactor. There are two ways to achieve kinematic similarity for scale-up (see Fig. 20): (i) maintain the same superficial velocities across the scales (and change the space velocity or the average residence time) and (ii) maintain the same average residence time or the space velocity (and change superficial velocities). Thermal similarity involves maintaining similar heat transfer characteristics. Small-scale reactors are generally operated under isothermal conditions. However, maintaining isothermal conditions in large-scaled reactors is often difficult and these are usually operated under

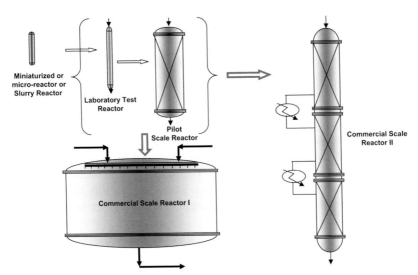

FIGURE 19 Various stages involved in the scale-up of trickle bed reactors.

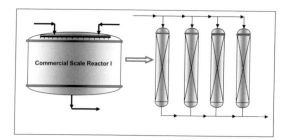

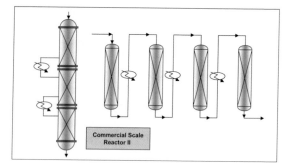

FIGURE 20 Methodologies to reduce the adverse effects of scale-up.

adiabatic conditions. When reactions are highly exothermic, consideration of thermal similarities is essential so that prevailing heat transfer and temperature field does not alter the hydrodynamics and therefore performance. In order to maintain the thermal similarity, additional inter-stage cooling or heating systems and addition of inerts are usually employed.

Influence of reactor scale on key design parameters and reactor performance is discussed in the following.

Reactor Parameters, Scale-Up, and Performance

Pressure Drop

The bed pressure drop is one of the important parameters while designing a commercial-scale reactor. In laboratory-scale reactors, pressure drop per unit length is negligible mainly due to smaller heights and higher porosity (~ 0.5). For lower D/d_p ratio (<5), porosity variation near wall is large. During the scale-up, change in particle diameter ($\sim 1-5$ times) is significantly lower than that in reactor diameter ($\sim 20-100$ times). Therefore, porosity variations near wall in large-scale reactors are usually negligible and average bed porosity is lower (~ 0.4).

Reactor pressure decreases along the length of the reactor. Possible reduction in density of fluids (due to pressure and temperature changes) and molar changes due to reaction may also influence pressure profile along the reactor length. Decrease in pressure along the length of the reactor can affect reaction rate due to decrease in gas solubility especially when gaseous reactants are limiting. In many commercial reactors, pressure drop across the bed is high and therefore in addition to operating pressure, decrease in pressure along the bed must be taken into account during scale-up. Funk et al. (1990) have shown that reaction rates increase with liquid flow rates up to a certain extent and then decrease mainly because of large pressure drop. If reaction is not intraparticle diffusion limited, one may use larger-sized particles and reduce bed pressure drop ($\Delta P/L \sim 1/d_p^2$).

Reactor Aspect Ratio

Aspect ratio of the scaled reactor needs to be designed based on geometric or kinematic similarities. Length to diameter ratio of commercial-scale reactors ($\sim 0.5-10$) is considerably different than pilot or bench-scale reactors ($\sim 5-100$). Salient features of tall and short reactors are listed in Table 2. These features can be useful for selection of appropriate aspect ratio for the case under consideration.

TABLE 2 Effect of Reactor Aspect Ratio on Key Reactor Characteristics

Reactors with High Aspect Ratio	Reactors with Low Aspect Ratio
• Higher gas—liquid superficial velocities	• Lower superficial velocities
• Approach plug flow behavior	• Higher deviation from plug flow
• Higher gas—liquid/liquid—solid mass transfer rates	• Lower gas—liquid/liquid—solid mass transfer rates
• Better radial liquid distribution	• Difficult to achieve radially uniform distribution
• Liquid redistribution over the length may be required	• Lower wetting efficiencies
• Higher bed wetting can be achieved	• Internal cooling/heating coils are suitable
• Outer heating jackets/intermediate cooling is possible/suitable	• Prone to non-uniform temperature in adiabatic operation
• Better control on temperature distribution within the bed	• Prone to hotspots and runaway conditions in the case of strong exothermic reactions and adiabatic operation
• Physical strength of catalyst particle is important	• Lower pressure drop/load on compressor
• Higher pressure drop/load on compressor	• Use of smaller particle size is permissible due to low gas—liquid velocities and lower pressure drop
• For diffusion-limited reactions, larger pressure drop across bed imposes limits on particle diameters which could be used	• Reactors can be operated in series
• Enhances foaming	• Suitable for foaming liquids
• Suitable for pulse/periodic flow operation	• Difficult to achieve/maintain pulse/periodic flow operation

Gas–Liquid Flow Rates

Maintaining similar gas and liquid superficial velocities in laboratory-scale and commercial-scale reactors is generally difficult. Typical gas–liquid flow rates in laboratory-scale, pilot-scale, and commercial-scale reactors are shown in Fig. 21. Most of the commercial-scale reactors are operated at higher throughputs. Due to high gas–liquid flow rate, performance of the commercial-scale reactor is considerably different from laboratory-scale reactor. Higher liquid flow rates in commercial-scale reactor increase liquid holdup, liquid–solid wetting, mass transfer rates, and to some extent heat transfer rates. Typical values of the gas–liquid flow rates for commercial-scale, pilot-scale, bench-scale reactors and micro-reactors are also listed in Table 3. Range of these parameters may vary depending upon the properties of the fluids and the reaction system. In typical hydrogenation and oxidation reactors, volume of material processed is considerably lower than the petroleum-refining processes. Superficial velocities used in petroleum processing reactors, however, can be considerably low due to large diameter of the reactors and requirement of complete conversion in the process. In typical hydrotreating trickle bed reactors, low liquid superficial velocity is desired for higher conversion. Therefore,

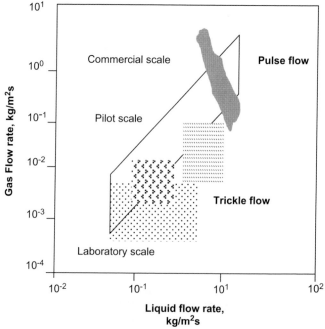

FIGURE 21 Typical gas–liquid flow ranges for different scales of trickle bed reactors (Satterfield, 1975).

TABLE 3 Typical Variation of Key Parameters with Different Scales of Trickle Bed Reactor

Parameters/Reactor scale	Commercial-Scale Reactor	Pilot-Scale Reactor	Bench-Scale Reactor	Micro-Reactor
Liquid flow rates ($kg/m^2/s$)	2.5–25	0.8–6	0.08–0.25	0.03–0.09
Gas flow rates ($kg/m^2/s$)	0.4–4	0.01–1	0.01–0.08	0.0001–0.05
Wetting efficiency	0.6–1	0.4–0.9	0.1–0.7	0.8–1
Liquid holdup	0.16–0.25	0.1–0.2	0.05–0.14	0.15–0.25
Gas–liquid mass transfer rates (s^{-1})	0.08–0.14		0.02–0.08	3–7
Liquid–solid mass transfer rates (s^{-1})	0.1–0.3		0.9–1.4	–
Length of bed (m)	16	1–4	0.3–1	0.008–0.5

wetting parameters in such reactors are considerably lower than the trickle bed reactors used for oxidation or hydrogenation reactions.

The simulated performance of typical laboratory-scale and commercial-scale hydrotreating reactors is shown in Fig. 22. In this case study, performance enhancement in commercial-scale reactor was found to be about 20–30% over the laboratory scale. This was mainly attributed to increased liquid holdup and mass transfer rates under commercial operating conditions. Uncertainty about the extent of wetting or maldistribution in large reactors is one of the key concerns in scale-up. The CFD models discussed in the previous chapter can provide useful insight and design guidelines for controlling and manipulating wetting and liquid distribution within the bed.

The ranges of parameters given in Table 3 are indicative. Trickle bed reactors are operated with a wide variety of conditions and these conditions differ considerably depending upon the type of application. For example, in typical oxidation reactions, higher gas–liquid mass transfer rates are desired for better performance of the reactor. Therefore, it is important to see how key parameters like interfacial area and mass transfer rates are changing with scale. For hydrocracking reactors, reactions are strongly exothermic, rather slow and often limited by liquid reactants. Therefore, for this case, heat transfer and adequate heat removal arrangement become far more crucial than the external mass transfer issues. Severe operating conditions ($T = 300-450\,°C$, $P = 100-250$ bar) alter the phase properties locally inside the reactor. Therefore, liquid

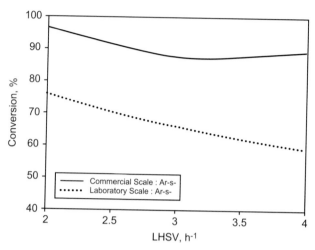

FIGURE 22 Comparison of performance of the commercial-scale and laboratory-scale hydro-treating reactors (Gunjal & Ranade, 2007). System: hydrotreating reactors; $P_{\text{lab-com}} = 4$ and 4.4 MPa; $\text{LHSV}_{\text{lab-com}} = 2.0$ and 3.0 h^{-1}; $(Q_G/Q_L)_{\text{lab-com}} = 200$ and 300 m^3/m^3; $y_{\text{H}_2\text{S}} = 1.4\%$).

holdup, wetting, extent of gas phase reactions, and mass transfer rates in commercial reactors can be considerably different from pilot-scale or bench-scale operation and not necessarily follow the trends shown in Table 3.

Most of the small-scale operations are carried out under isothermal conditions and uniform temperature in the bed. However, most of the commercial-scale reactors are operated under adiabatic conditions in which variation of local temperature affects the performance of reactor considerably. In fact, in some operations intermediate cooling or gas quenching is necessary to maintain the operating temperatures in the desired range, to avoid hotspots and runaway conditions. Many commercial-scale trickle bed reactors are multistage reactors typically 1−4 m in diameter and 4−6 m in length arranged in 3−4 stages. Intermediate cooling and liquid redistribution are possible using such arrangements. Despite these arrangements, formation of local hotspots is possible due to local liquid maldistribution effects. Because of liquid maldistribution, liquid flow rates are different in different regions of trickle bed reactors. The regions with low liquid flow rates may experience rise in temperature (Jaffe, 1976). An example of such local variation in reactor temperature for the hydrocracking reactor is shown in Fig. 23. Sharp peaks in temperature along the length of the bed indicate formation of hotspots, which may cause damage to the catalyst due to overheating. These results indicate that the formation of local hotspots is independent of position of interstage cooling and liquid maldistribution may be the most probable reason for formation of local hotspots.

Detection of low liquid flow regions or locations of hotspot is difficult under such high temperature and pressure conditions. Therefore, it is important to gain the information on extent of maximum adiabatic temperature rise, heat

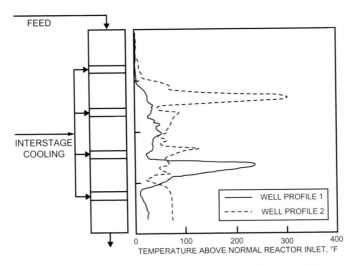

FIGURE 23 Axial temperature profile in commercial reactor is independent of interstage cooling (Jaffe, 1976). System: hydrocracking of oil.

dissipation rate, and extent of high temperature region. Such quantification is possible by using mathematical models based on mixing tank network where artificial flow obstacle produces low flow regions. Possibility of hotspots and sensitivity of these with key design and operating parameters can be estimated using the reaction kinetics and appropriate reactor models. Several assumptions associated with the flow simplification and bed characterization can be eliminated by making use of CFD-based model. The detailed three-dimensional CFD models can capture liquid maldistribution as well as axial and radial mixing of gas and liquid phases within the reactor. Such models are useful for providing the guidelines required for reducing the extent of hotspots, severity of the problem, and the risk associated with rising of temperature to the adiabatic flash point. The overall approach of engineering of trickle bed reactors is discussed in the following section. Some examples and case studies are discussed in Chapter 6.

ENGINEERING OF TRICKLE BED REACTORS

Reactor engineering of trickle bed involves all the activities necessary to evolve best-possible hardware and operating protocols for carrying out the desired transformation of raw materials (or reactants) to value-added products. Naturally, it requires expertise from various fields ranging from thermodynamics, chemistry and catalysis to reaction engineering, fluid dynamics, mixing, and heat and mass transfer. Reactor engineer has to interact with chemists to understand the basic chemistry and peculiarities of the catalyst. Based on such understanding and proposed performance targets, reactor engineer has to conceive a suitable reactor hardware and operating protocols. Typical "wish list" of a reactor engineer for the new process could be as follows:

- Stable operation within technologically feasible region
- Intrinsically safe operation
- Environmentally acceptable
- Maximum possible conversion of the feedstocks
- Maximum selectivity of reaction to the desired products
- Lowest capital and operating costs

Typical "wish list" for enhancing performance of the existing reactor technology could be as follows:

- More throughput per unit volume
- Improved selectivity and better quality product
- Safer operation
- Reduced energy consumption
- More environment friendly operation

The next step is to translate the wish list into a quantitative form and establish a relationship between items in the wish list and reactor hardware and operating protocols. The reactor engineer's task is to design and tailor the reactor hardware and operating protocols to realize the wish list. Several activities are involved in this process. It may often turn out that some of the items in the wish list require contradictory options of hardware and operation. In such a case, a careful analysis of different items in wish list must be made to assign priorities. Operability, stability, and environmental constraints often get precedence over conversion and selectivity when such conflicting requirements arise.

The first step, in engineering of trickle beds, involves quantification of influence of reactant flow rate and operating temperature on performance of the reactor (conversion, selectivity, stability, and so on). The first step in such analysis is formulating a mathematical framework to describe the rate (and mechanism) by which one chemical species is converted into another in absence of any transport limitations (chemical kinetics). Kinetics of reactions need to be determined from experimental measurements. Measuring the rate of chemical reactions in the laboratory is itself a specialized branch of chemical engineering. More information about laboratory reactors used for obtaining intrinsic kinetics can be found in reviews/textbooks like Chaudhari, Shah, and Foster (1986), Doraiswamy and Sharma (1984), Levenspiel (1972), Ramachandran and Chaudhari (1983), and Smith (1970).

Once the intrinsic kinetics is available, the effective rate analysis can be carried out using the methods discussed in Chapter 3. The conventional reaction engineering models for trickle bed reactors can be developed based on this analysis which may help in establishing a relationship among operating conditions (flow rate, mode of operation, pressure, composition),' reactor hardware (reactor volume and reactor configuration) and reactor performance.

As discussed in Chapter 3, in these models, some assumptions are made regarding the flow and mixing of various species in the reactor, instead of solving the detailed fluid dynamics equations. These models therefore cannot directly relate the reactor hardware with reactor performance. These models are however computationally much less demanding than the CFD-based models and can give a quick understanding of the overall behavior of the reactor. These models can be used to identify the important parameters/issues, which may require further study. Of course, the class of conventional chemical reaction engineering models itself contains a variety of models. It will be useful to distinguish between "learning" models and "design" models at this stage.

"Learning" models are developed to understand the basic concepts and to obtain specific information about some unknown processes. The results obtainable from such models may not directly lead to design information but are generally useful to take appropriate engineering decisions. "Design" models, on the other hand, yield information or results, which can be directly used for reactor design and engineering. It is first necessary to develop some design models to estimate reactor sizing and to evolve preliminary reactor configuration. Several "learning" models can then be developed to understand various reactor engineering issues like:

- Start-up and shut-down dynamics
- Multiplicity and stability of thermo-chemical processes occurring in the reactor
- Sensitivity of reactor performance with respect to mixing and RTD
- Selectivity and by-product formation

After establishing such understanding and analysis of the reaction system, the next most important question facing the reactor engineer is to evaluate consequences of the assumptions involved in such models for estimating the behavior of actual reactor. Questions like operability and stability of the flow regime need to be answered. Mixing, wetting as well as heat and mass transfer are intimately related to flow regimes and distributors of gas and liquid. The mixing in actual reactor may significantly deviate from that assumed for the reaction engineering models. This deviation can be caused by say channeling of fluid or by formation of stagnant regions within the reactor. RTD data and analysis is useful to get the bounds and limiting solutions of reactor perfor-mance. It must be remembered that more than one model may fit the observed RTD data. A general philosophy is to select the simplest possible model, which adequately represents the physical phenomena occurring in the actual reactor. The RTD data and models are useful to understand and quantify possible influence of static liquid holdup on reactor performance and product quality. The brief review of distributors given in *Reactor Performance* along with the references cited therein will be useful to finalize selection of distributors. It may be noted that even the small-scale hardware details like design of feed nozzles or weirs and reactor internals (cooling coils, baffles etc.) may have dramatic

influence on reactor performance. The issues of scale-up and scale-down may
have to be resolved by following an iterative methodology. Engineering crea-
tivity, experience, and accumulated empirical information is generally used to
evolve reactor configuration and designs. The discussion in Chapter 3 as well as
in *Reactor Performance* and *Trickle Bed Reactor Design and Scale-Up* will be
useful to implement these steps.

The understanding gained by development and implementation of these
steps will be helpful to identify the needs for developing more sophisticated
simulation models/data for establishing the desired trickle bed reactor design. It
is important at this stage to identify gaps between available knowledge and that
required for fulfilling the "wish list." The identified gaps can then be bridged by
carrying out experiments in the laboratory and/or pilot plant(s) and by developing more comprehensive fluid dynamic models. Despite the advantages,
conventional chemical reaction engineering models are not directly useful for
understanding the influence of reactor hardware on reactor performance. For
example, how the design of distributor affects the radial distribution of liquid in
trickle bed and thereby the reactor performance will be difficult to predict
without developing the detailed fluid dynamic model of the reactor or without
carrying out experiments on a scale model. The CFD-based approach will make
valuable contributions at this stage by providing the required insight, by
helping to devise right kind of experiments, and by allowing screening of
alternative configurations and tools for extrapolation and scale-up. The methodology and models discussed in Chapter 4 will be useful for this purpose.

The whole process of reactor engineering is not sequential! All the steps
interact with and influence each other. The results obtained in the laboratory
experiments on hydrodynamics and RTD or from the computational flow model
may demand changes and revisions in the earlier analysis and the whole process
is iterated until the satisfactory solution emerges. In this book, we have tried to
provide adequate details of models and methodologies useful to carry out
different components of reactor engineering of trickle beds. Some applications
of this engineering methodology are discussed in the next chapter (Chapter 6).

SUMMARY

In engineering practice, reactor engineering solution is composed of three main
steps: (i) analysis of transport and reaction rate parameters; (ii) identifying
uncertainties associated with design parameters; and (iii) resolving process
complexities and conflicting demands. In trickle bed reactors, interaction
between reaction kinetics and hydrodynamics is often complex. This chapter
complements discussion covered in Chapters 2–4 and provides a methodology
for engineering of trickle bed reactors.

Key practical aspects involving rate analysis including hysteresis and
multiplicity, particle and bed properties (particle size/shape/orientation),
distributor effects, liquid phase maldistribution, and mixing of phases along the

length of the column are discussed. Issues related to selection of reactor aspect ratio and issues pertaining to scale-up and scale-down are discussed. An overall methodology combining experiments and computational modeling is presented at the end. Some examples of engineering of trickle beds as well as a brief review of recent developments and path forward in trickle bed reactors are discussed in the following chapter. It is evident from current state of the art that in spite of considerable advances in our understanding of different phenomena occurring in trickle beds, it is futile to attempt a single comprehensive model for a trickle bed reactor incorporating all the processes. It is therefore recommended to use a suite of models comprising many 'learning' and 'design' models. This modeling efforts need to be complemented by appropriately designed experiments of three categories: learning, calibration and validation. Such a composite approach based on the methodology discussed here will be useful for engineering of trickle bed reactors.

REFERENCES

Al-Dahhan, M. H., Wu, Y. X., & Dudukovic, M. P. (1995). Reproducible technique for packing laboratory-scale trickle-bed reactors with a mixture of catalyst and fines. *Industrial and Engineering Chemistry Research, 34*, 741.

Ballard, J. H., & Hines, J. E. (1965). Vapor Liquid Distribution Method and Apparatus for the Conversion of Hydrocarbons. *U.S. Patent No 3,218,249.*

Boelhouwer, J. G., Piepers, H. W., & Drinkenburg, A. A. H. (2002). Liquid-induced pulsing flow in trickle-bed reactors. *Chemical Engineering Science, 57*, 3387–3399.

Borda, M., Gabitto, J. F., & Lemcoff, N. O. (1987). Radial liquid distribution in a trickle bed reactor. *Chemical Engineering Communications, 60*, 243–252.

Bruijn, A.de, Naka, I., & Sonnemans, J. W. M. (1981). Effect of the noncylindrical shape of extrudates on the hydrodesulfurization of oil fractions. *Industrial and Engineering Chemistry Process Design and Development, 20*, 40–45.

Cassanello, M. C., Martinez, O. M., & Cukierman, A. L. (1992). Effect of the liquid axial dispersion on the behavior of fixed bed three phase reactors. *Chemical Engineering Science, 47*(13–14), 3331–3338.

Castellari, A. T., & Haure, P. M. (1995). Experimental study of the periodic operation of a trickle-bed reactor. *AIChE Journal, 41*, 1593–1597.

Chaudhari, R. V., Jaganathan, R., Mathew, S. P., Julcour, C., & Delmas, H. (2002). Hydrogenation of 1,5,9-cyclododecatriene in fixed-bed reactors: down- vs. upflow modes. *AIChE Journal, 48*(1), 110–125.

Chaudhari, R. V., Shah, Y. T., & Foster, N. R. (1986). Novel gas–liquid–solid catalytic reactors. *Catalysis Reviews – Science and Engineering, 28*, 431.

Chou, T. S. (1984). Liquid distribution in a trickle bed with redistribution screens placed in the column. *Industrial and Engineering Chemistry Process Design and Development, 23*, 501.

Crine, M., & L'Homme, G. A. (1984). In L. K. Doraiswamy (Ed.), *Recent advances in the engineering analysis of chemically reacting systems* (pp. 430). New York: Wiley.

Crine, M., Marchot, P., & L'Homme, G. (1981). Liquid flow maldistributions in trickle-bed reactors. *Chemical Engineering Communications, 7*, 377–388.

Doraiswamy, L. K., & Sharma, M. M. (1984). *Heterogeneous Reactions: Analysis, Examples and Reactor Design, Vol 2.* New York: John Wiley.

Fox, R. (1987). On the liquid flow distribution in trickle-bed reactor. *Industrial and Engineering Chemistry Research, 26*, 2413–2419.

Funk, G. A., Harold, M. P., & Ng, K. M. (1990). A novel model for reaction in trickle beds with flow maldistribution. *Industrial and Engineering Chemistry Research, 29*, 738–748.

Gabarain, L., Castellari, A. T., Cechini, J., Tobolski, A., & Haure, P. M. (1997). Analysis of rate enhancement in a periodically operated trickle-bed reactor. *AIChE Journal, 43*, 166–177.

Gamborg, M. M., & Jensen, B. N. (1999). Two-Phase Downflow Liquid Distribution Device. *U.S. Patent No. 5,942,162.*

Goto, S., & Smith, J. M. (1975). Trickle-bed reactor performance. Part-I: holdup and mass transfer effects. *AIChE Journal, 21*, 706.

Gulijk, C. V. (1998). Using computational fluid dynamics to calculate transversal dispersion in a structured packed bed. *Computers and Chemical Engineering, 22*, 5767–5770.

Gunjal, P. R., Kashid, M. N., Ranade, V. V., & Chaudhari, R. V. (2005). Hydrodynamics of trickle-bed reactors: experiments and CFD modeling. *Industrial and Engineering Chemistry Research., 44*, 6278–6294.

Gunjal, P. R., & Ranade, V. V. (2007). Modeling of laboratory and commercial scale hydro-processing reactors using CFD. *Chemical Engineering Science, 62*, 5512–5526.

Gunjal, P. R., Ranade, V. V., & Chaudhari, R. V. (2003). Liquid-phase residence time distribution in trickle bed reactors: experiments and CFD simulations. *The Canadian Journal of Chemical Engineering, 81*, 821.

Hanika, J., Lukjanov, B. N., Kirillov, V. A., & Stanek, V. (1986). Hydrogenation of 1,5-cyclo-octadiene in a trickle bed reactor accompanied by phase transition. *Chemical Engineering Communications, 40*, 183.

Hanika, J., Sporka, K., Ruzicka, V., & Krausova, J. (1975). Qualitative observations of heat and mass transfer effects on the behaviour of a trickle bed reactor. *Chemical Engineering Communications, 2*, 19.

Hanika, J., Sporka, K., Ruzicka, V., & Pistek, R. (1977). Dynamic behavior of an adiabatic trickle bed reactor. *Chemical Engineering Science, 32*, 525.

Hanratty, P. J., & Dudukovic, M. P. (1992). Detection of flow maldistribution in trickle-bed reactors via tracers. *Chemical Engineering Science, 47*, 3003–3014.

Haure, P., Hudgins, R. R., & Silveston, P. L. (1989). Periodic operation of a trickle-bed reactor. *AICh E Journal, 35*, 1437–1444.

Herskowitz, M., & Mosseri, S. (1983). Global rates of reaction in trickle-bed reactors: effects of gas and liquid flow rates. *Industrial and Engineering Chemistry Fundamentals, 22*, 4–6.

Herskowitz, M., & Smith, J. M. (1978). Liquid distribution in trickle bed reactors. *AIChE Journal, 24*, 439.

Hessari, F. A., & Bhatia, S. K. (1997). Hysteresis and phase transitions in a single partially internally wetted catalyst pellet: thermo-gravimetric studies. *Chemical Engineering Communications, 157*, 109–133.

Higler, A. P., Krishna, R., Ellenberger, J., & Taylor, R. (1999). Counter-current operation of a structured catalytically packed-bed reactor: liquid phase mixing and mass transfer. *Chemical Engineering Science, 54*(21), 5145–5152.

Hochman, J., & Effron, E. (1969). Two-phase cocurrent downflow in packed beds. *Industrial and Engineering Chemistry Fundamentals, 8*(1), 63–69.

Jaffe, S. B. (1976). Hot spot simulation in commercial hydrogenation processes. *Industrial and Engineering Chemistry Process Design and Development, 15*(3), 410–416.

Jiang, Y., Khadilkar, M. R., Al-Dahhan, M. H., & Dudukovic, M. P. (1999). Two-phase flow distribution in 2D trickle-bed reactors. *Chemical Engineering Science, 54*, 2409–2419.

Jiang, Y., Khadilkar, M. R., Al-Dahhan, M. H., & Dudukovic, M. P. (2001). CFD modeling of multiphase flow distribution in catalytic packed bed reactors: scale down issues. *Catalysis Today, 66*, 209–218.

Julcour, C., Jaganathan, R., Chaudhari, R. V., Wilhelm, A. M., & Delmas, H. (2001). Selective hydrogenation of 1,5,9-cyclododecatriene in up- and down-flow fixed-bed reactors: experimental observations and modeling. *Chemical Engineering Science, 56*, 557–564.

Khadilkar, M. R., Al-Dahhan, M. H., & Dudukovic, M. P. (1999). Parametric study of unsteady-state flow modulation in trickle-bed reactors. *Chemical Engineering Science, 54*, 2585–2595.

Khadilkar, M. R., Wu, X., Al-Dahhan, M. H., Dudukovic, M. P., & Colakyan, M. (1996). Comparison of trickle-bed and upflow reactor performance at high pressure: model predictions and experimental observations. *Chemical Engineering Science, 51*(10), 2139–2148.

Kheshgi, H. S., Reyes, S. C., Hu, R., & Ho, T. C. (1992). Phase transition and steady-state multiplicity in a trickle-bed reactor. *Chemical Engineering Science, 47*, 1771.

Kim, D. H., & Kim, Y. G. (1981). An experimental study of multiple steady states in a porous catalyst due to phase transition. *Journal of Chemical Engineering of Japan, 14*, 311–317.

Kireenkov, V. V., Shigarov, A. B., Kuzin, N. A., Bocharov, A. A., & Kirillov, V. A. (2006). Effect of the nonuniformity of the inlet liquid distribution on the trickle-bed reactor output in an exothermic reaction accompanied by evaporation. *Theoretical Foundations of Chemical Engineering, 40*(5), 472–482.

Kirillov, V. A., & Koptyug, I. V. (2005). Critical phenomena in trickle-bed reactors. *Industrial and Engineering Chemistry Research, 44*, 9727–9738.

Klerk, A.de (2003). Voidage variation in packed beds at small column to particle diameter ratio. *AIChE Journal, 49*(8), 2022–2029.

Lange, R., Hanika, J., Stradiotto, D., Hudgins, R. R., & Silveston, P. L. (1994). Investigations of periodically operated trickle-bed reactors. *Chemical Engineering Science, 49*(24B), 5615–5621.

Larachi, F., Laurent, A., Midoux N., & Wild, G. (1991), Experimental study of a trickle-bed reactor Operating at high pressure: two-phase pressure drop and liquid saturation, *Chemical Engineering Science, vol. 46*, no. 5,6, pp. L233–1246.

Lee, J. K., Hudgins, R. R., & Silveston, P. L. (1995). Cycled trickle-bed reactor for SO$_2$ oxidation. *Chemical Engineering Science, 50*, 2523–2530.

Liu, G., Mi, Z., Wang, L., Zhang, X., & Zhang, S. (2006). Hydrogenation of dicyclopentadiene into *endo*-tetrahydrodicyclopentadiene in trickle-bed reactor: experiments and modeling. *Industrial and Engineering Chemistry Research, 45*, 8807–8814.

Lutran, P. G., Ng, K. M., & Delikat, E. P. (1991). Liquid distribution in trickle beds: an experimental study using computer-assisted tomography. *Industrial and Engineering Chemistry Research, 30*, 1270–1280.

Marcandelli, C., Lamine, A. S., Bernard, J. R., & Wild, G. (2000). Liquid distribution in trickle-bed reactor. *Oil & Gas Science and Technology – 'Revue d'IFP Energies nouvelles'. IFP, 55* (4), 407–415.

McManus, R. L., Funk, G. A., Harold, M. P., & Ng, K. M. (1993). Experimental study of reaction in trickle-bed reactors with liquid maldistribution. *Industrial and Engineering Chemistry Research, 32*, 510–514.

Mears, D. E. (1971). The role of axial dispersion in trickle flow laboratory reactors,. *Chemical Engineering Science, 26*, 1361.

Metaxas, K. C., & Papayannakos, N. G. (2006). Kinetics and mass transfer of benzene hydrogenation in a trickle-bed reactor. *Industrial and Engineering Chemistry Research, 45*, 7110–7119.

Møller, L. B., Halken, C., Hansen, J. A., & Bartholdy, J. (1996). Liquid and gas distribution in trickle-bed reactors. *Ind. Eng. Chem. Res., 35*, 926–930.

Montagna, A. A., & Shah, Y. T. (1975). The role of liquid holdup, effective catalyst wetting, and back-mixing on the performance of a trickle bed reactor for residue hydrodesulfurization. *Industrial and Engineering Chemistry Process Design and Development, 14*(4), 479–483.

Nemec, D., & Levec, J. (2005). Flow through packed bed reactors: 1. Single phase flow. *Chemical Engineering Science, 60*, 6947–6957.

Oliveros, G., & Smith, J. M. (1982). Dynamic studies of dispersion and channeling in fixed beds. *AIChE Journal, 28*, 751–759.

Rajashekharam, M. V., Jaganathan, R., & Chaudhari, R. V. (1998). A trickle-bed reactor model for hydrogenation of 2,4 dinitrotoluene: experimental verification, *Chemical Engineering Science, 53*(4), 787–805.

Ramachandran, P. A., & Chaudhari, R. V. (1983). *Three Phase Catalytic Reactors*. New York, USA: Gordon and Breach.

Ramachandran, P. A., & Smith, J. M. (1978). Dynamic behavior of trickle bed reactor. *Chemical Engineering Science, 34*, 75–91.

Ravindra, P. V., Rao, D. P., & Rao, M. S. (1997). A model for the oxidation of sulfur dioxide in a trickle-bed reactor,. *Industrial and Engineering Chemistry Research, 36*, 5125–5132.

Riopelle, J. E., & Scarsdale, N. Y. (1967). Bed Reactor with Quench Deck. *U.S. Patent No 3,353,924.*

Saroha, A. K., Nigam, K. D. P., Saxena, A. K., & Kapoor, V. K. (1998). Liquid distribution in trickle-bed reactors. *AIChE Journal, 44*, 2044–2052.

Satterfield, C. N. (1975). Trickle bed reactors. *AIChE Journal, 21*(2), 209–228.

Schwartz, J. G., Weger, E., & Dudukovic, M. P. (1976). A new tracer method for the determination of liquid–solid contacting efficiency in trickle bed reactors. *AIChE Journal, 22*, 953.

Sederman, A. J., & Gladden, L. F. (2001). Magnetic resonance imaging as a quantitative probe of gas–liquid distribution and wetting efficiency in trickle-bed reactors. *Chemical Engineering Science, 56*, 2615–2628.

Shigarov, A. B., Kuzin, N. A., & Kirillov, V. A. (2002). Phase disequilibrium in the course of an exothermic reaction accompanied by liquid evaporation in a catalytic trickle-bed reactor. *Theoretical Foundations of Chemical Engineering, 36*, 159–165.

Silveston, P. L., & Hanika, J. (2002). Challenges for the periodic operation of trickle-bed catalytic reactors. *Chemical Engineering Science, 57*, 3373–3385.

Smith, J.M., 1970, *Chemical Engineering Kinetics*, 2 Edition, McGraw-Hill Book.

Stegeman, D., van Rooijen, F. E., Kamperman, A. A., Weijer, S., & Westerterp, K. R. (1996). Residence time distribution in the liquid phase in a cocurrent gas–liquid trickle bed reactor. *Industrial and Engineering Chemistry Research, 35*, 378–385.

Sun, C. G., Yin, F. H., Afacan, A., Nandakumar, K., & Chuang, K. T. (2000). Modelling and simulation of flow maldistribution in random packed columns with gas–liquid countercurrent flow. *Chemical Engineering Research and Design, 78*(3), 378–388.

Sylvester, W. D., & Pitayagulsarn, P. (1975). Radial liquid distribution in cocurrent two phase downflow in packed beds. *The Canadian Journal of Chemical Engineering, 53*, 599.

Van Swaaij, W. P. M., Charpenter, J. C., & Villermaux, J. (1969). Residence time distribution in the liquid phase of trickle flow in packed columns. *Chemical Engineering Science, 24*, 1083–1095.

Watson, P. C., & Harold, M. P. (1993). Dynamic effects of vaporization with exothermic reaction in a porous catalytic pellet. *AIChE Journal, 39*, 989–1006.

Weekman, V. W., Jr., & Myers, J. (1964). Fluid flow characteristics of cocurrent gas–liquid flow in packed beds. *AIChE Journal, 10*, 951.

Chapter 6

Applications and Recent Developments

Engineering refers to the practice of organizing the design, construction and operation of any artifice which transforms the physical world around us to meet some recognized need.

GFC Rogers

INTRODUCTION

Trickle bed reactors are extensively used in chemical and associated industries such as petroleum, petrochemical, oil and gas, mineral and coal industries, pharmaceuticals, fine and specialty chemicals, biochemicals, and waste treatment. Applications may vary considerably from industry to industry and may include cracking of large organic molecules into useful desired products, upgrading petroleum feedstock, conversion of unsaturated organics into saturated products, conversion of coal-derived products, conversion of gaseous reactants into fuels, hydrodeoxygenation (HDO) of bio-oils (upgrading), polymerization of monomers for a variety of commodity applications, stripping of unwanted chemicals, purification of feedstocks, manufacturing pharmaceutical and their intermediates, aromatization and dearomatization of organic materials used typically in specialty chemicals, oxidation, demetallization, denitrification, dewaxing, desulfurization, removal of pollutants, and wastewater treatment. Trickle bed reactors are preferred for large volume processes. Scales and key operational characteristics of trickle bed reactors used in different applications cover a broad range. In this chapter, key design issues are illustrated with the help of some examples.

As discussed earlier, trickle bed reactors offer several advantages like simplicity in operation (without any moving part or catalyst separation unit), high catalyst loading per unit volume, and low capital and operating costs. It also has some inherent drawbacks like poor external and intraparticle heat transfer rates, significant intraparticle diffusion limitations, and susceptibility to liquid maldistribution. Therefore, design of reactors for a particular application often requires balance among different competing requirements. For example, reduction in operating cost (by reducing pressure drop) is possible by

Trickle Bed Reactors. DOI: 10.1016/B978-0-444-52738-7.10003-8

(212)   Trickle Bed Reactors

using larger-sized particles, however, larger particles lead to lower effectiveness factors because of intraparticle diffusion limitations. Trickle bed operation is simple at the expense of fewer degrees of freedom available to engineers for manipulating its performance. There is limited scope to manipulate local conditions within the bed such as local hotspots (which may lead to sintering or runaway) and use of catalysts with rapid deactivation. The liquid space velocity and gas flow rates used in laboratory and commercial trickle bed reactors may vary by about 10-fold which can lead to different flow regimes in these two scales. Moreover, in one case the external mass transfer may be limiting while in another it may be unimportant. These issues can seriously affect the scale-up and design strategy of reactors. The reactor models therefore have to be carefully coupled with diagnostic analysis in each case to assess the likely rate-limiting processes on each scale. The key issues, models, and methodologies discussed in Chapters 2–5 are useful to the reactor engineer to fully realize advantages of trickle bed reactors in practice and to evolve strategies to mitigate implications of some of the inherent disadvantages of trickle bed reactors. Attempt is made here to illustrate this with the help of a few examples.

In recent years, other variants of trickle bed reactors like monolith reactors or micro-trickle bed reactors have been evolved which may minimize some of the inherent disadvantages of the trickle bed reactors while significantly improving the performance. These recent developments are briefly discussed in the second part of this chapter. Aspects related to the reaction engineering, design, and operation of monolith as well as micro-trickle bed reactors are also discussed. A closure section recapitulates the approach discussed in this book for reactor engineering of trickle beds. Some comments on future trends are also included.

EXAMPLES OF TRICKLE BED REACTOR APPLICATIONS

In this section, some examples taken from our research and consulting experience are discussed. Naturally there are constraints imposed by the confidentiality obligations. Here the purpose is to discuss the approach and the methodology rather than discussing the specific details of implementation.

Hydrogenation Reactions

Hydrogenation reactions are important in many sectors of chemical process industries ranging from petroleum to pharmaceuticals. In these processes unsaturated organic compounds are contacted with hydrogen on the active catalyst surface to make varying degrees of saturated organic compounds. Commonly used catalysts for hydrogenation reactions include noble metals like Palladium, Platinum, Ruthenium, Rhodium, and Nickel or combination of these supported on silica, alumina, or activated carbon, etc., or the non-noble metal catalysts consisting of Cobalt, Molybdenum, Copper and their

TABLE 1 Typical Hydrogenation Reactions Carried out in Trickle Bed Reactors

Hydrogenation Reaction	Other Details
Benzaldehyde	Ni/Al$_2$O$_3$, Pt/C
Cinnamaldehyde	Pt/C, Ir/C
γ-Butyrolactone	Ni/Al$_2$O$_3$ GLS
Styrene/octene, toluene	Ni/Al$_2$O$_3$, Pd/Al$_2$O$_3$
α-Methyl styrene	Ni/Al$_2$O$_3$, Ru
Benzene, dicyclopentadiene, 1,5-cyclooctadiene, aniline	
Xylose	Pd/Al$_2$O$_3$ GLS batch − continuous
2-Butyne-1,4 diol	Ni on C, continuous − cocurrent
Hydrogenation of caprolactone	Cu, continuous − cocurrent
Hydrogenation of glucose, sorbitol, acetophenone, glycerol, 2,4-dinitrotoluene, nitrobenzene	Ru/C
Hydrogenation of coal liquefaction extracts	Ni−Mo/Al$_2$O$_3$
Hydrocracking of heavy oil, paraffin, naphthene	Pd/Al$_2$O$_3$
Hydrotreating reactions, for example, hydrogenation mono−diaromatic hydrocarbon, polyhydrocarbon	
Hydrogenation of vegetable oils	Raney Ni
Pyrolysis of gasoline	Pd, Pt on Al$_2$O$_3$
Hydrogenation of itaconic acid	Pd/C

combinations on appropriate supports. Some examples of hydrogenation reactions used in petrochemical and fine/pharmaceutical industries are listed in Table 1. Hydrogenation reactions practiced in trickle bed reactors vary widely with respect to their reaction engineering aspects depending on the complexities of chemistry, reaction kinetics, deactivation, exothermicity, and scales of operation. For example, most of the hydrogenation reactions using supported noble metal catalysts are very rapid and highly exothermic leading to severe mass transfer limitations (gas to liquid). For such cases, the understanding of the flow regimes and hydrodynamic behavior is crucial in scale-up and design.

Similarly, on a catalyst particle-scale, intraparticle heat and mass transfer limitations can lead to local hotspots, incomplete wetting of catalyst (due to evaporation of volatile reactant/solvent), deactivation of catalysts, and ineffective utilization of the expensive catalysts. A more complex phenomenon of multiplicity of steady states, hysteresis, and oscillations is observed in many such cases representing rapid reactions coupled with exothermic reactions. For such cases, detailed reaction engineering models coupled with computational fluid dynamics (CFD) are most useful to represent reactor performance independent of the scale of operation. In another class of hydrogenation reactions, involved in hydroprocessing (hydrodesulfurization (HDS) or hydrodenitrogenation (HDN)) of petroleum oils, the reactions are relatively slower and hence very large volume reactors are used (also due to larger scales of production). Here, very often due to high pressure of hydrogen, the concentration of hydrogen is in large excess compared to the S or N-containing impurities present as liquid phase substrates, thus leading to the liquid reactant being a limiting reactant. The intraparticle diffusion of these bulky molecules is also very slow compared to that of dissolved hydrogen, leading to conditions where the liquid reactant/intermediates become limiting, followed by their highly complex diffusion within the porous catalysts. Such processes with lower overall productivity per unit weight of catalyst demand operation at lower liquid velocities to achieve complete conversion (complete removal of impurities) imposing incomplete wetting of the catalyst particles. The applications of hydrogenation reactions in fine chemicals and specialty products involve complexities due to thermal stability of the products and hence demand precise control of temperatures. Though conventionally batch slurry reactors are used in these processes due to smaller scale of operation compared to the commodity processes, trickle beds may prove to be advantageous in many of these processes considering safety and better control of selectivity of the products. Thus, hydrogenation reactions represent such a wide variety that all the important issues of reaction engineering of trickle bed reactors can be illustrated with appropriate models.

Hydrogenation reactions in different types of processes employ trickle bed reactors, which include, primarily, hydroprocessing (hydrodesulfurization, hydrocracking, hydrodenitrogenation, etc.) of petroleum oils (Ramachandran & Chaudhari, 1983) and hydrogenation of fine chemicals and pharmaceuticals (Mills, Ramachandran, & Chaudhari, 1992). The design requirements in these processes are widely different and in many cases, empirical pseudo-homogeneous models of the trickle bed reactors are used for design. The current state of development of detailed reactor models will allow a better insight into design and scale-up issues of these processes.

Several examples of using single reactions with first-order kinetics have been previously studied experimentally to understand the rate-limiting steps and partial wetting effects (Mills & Dudukovic, 1979; Ramachandran & Chaudhari, 1983; Satterfield, 1970). The kinetics of hydrogenation of α-methyl styrene (AMS) using supported Pd catalysts is first order in hydrogen and zero order in AMS. Mills and Dudukovic (1984) used this reaction system to show contacting efficiency in a trickle bed reactor considering partial external wetting of catalyst pellets with finite mass transfer resistances at the actively and inactively wetted catalyst surfaces. Typical results showing comparison of experimental and predicted conversions are shown in Fig. 1 for two different solvents, which show that the contacting efficiency decreases with increase in conversion of AMS. The contacting efficiency is a function of variety of factors as discussed in Chapter 2 and physical properties of solvent is one of the important parameters. In this particular case, the contacting efficiency of one of the solvents (hexane) decreases more rapidly than the other solvent (cyclo-hexane) as liquid flow rate decreases. This results into larger sensitivity of the AMS conversions to the relative liquid flow rate with hexane (as a solvent) than with the cyclohexane. The reaction engineering model developed by Mills and Dudukovic (1984) captures these trends quite adequately as seen from Fig. 1.

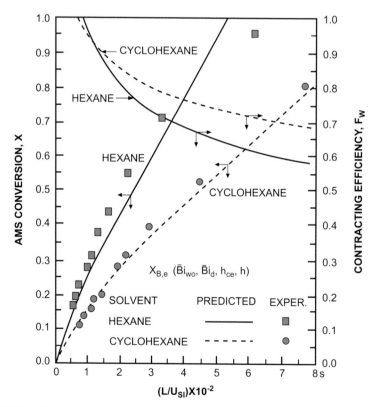

FIGURE 1 Experimental and model predicted conversion in AMS hydrogenation in trickle bed reactor (adapted from Mills & Dudukovic, 1983).

Trickle Bed Reactors

Hydrogenation of AMS has also been often used to illustrate evaluation of external and intraparticle mass transfer resistances based on experimental data (Ramachandran & Chaudhari, 1983). Couple of other examples with complex reactions are discussed in the following.

Hydrogenation of Acetophenone: Multistep Isothermal Reactions

Hydrogenation of acetophenone (ACPH) is an example involving several series and parallel catalytic reactions as shown in Eq. (1). Out of the various products shown in Eq. (1), 1-phenylethanol (1-PHET) and 1-cyclo-hexylethanol (CHET) are industrially important products, used as intermediates in pharmaceutical (e.g., ibuprofen) and perfumery industries. The first step in designing an industrial trickle bed reactor for such a complex system is to conduct experiments on laboratory scale, characterize time scales of different reactions, and develop intrinsic kinetic models (as described in Chapter 3). As an illustration of this, some results of Mathew (2001) are discussed here.

$$(1)$$

where, ACPH: acetophenone, PHET: phenylethanol, ETBE: ethylbenzene, CHMK: cyclohexyl methyl ketone, CHET: 1-cyclohexylethanol, and ETCH: ethyl cyclohexane.

A schematic of the laboratory reactor set-up used by Mathew (2001) and other relevant information about their experiments are included in Fig. 2. Overall hydrogenation involves two parallel reactions, i.e., (i) hydrogenation of the aromatic ring and (ii) hydrogenation of ketonic group (see Eq. (2)). Thus, formation of cyclohexylethanol can be either from the first hydrogenation of ketonic group or from the first hydrogenation of aromatic ring. Application of rate analysis methodologies discussed in Chapter 3 indicated that the former step was found to be faster than the latter (Mathew, 2001). Their study also revealed that the rates of formation of ETCH and CHET via CHMK were

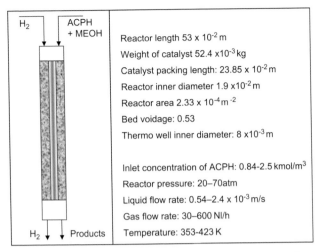

FIGURE 2 Schematic of laboratory set-up and other relevant parameters of experiments of Mathew (2001).

negligible compared to the other products. Therefore, following simplified form of reaction scheme was considered for ACPH hydrogenation:

$$
\begin{array}{ccccc}
\text{ACPH} & \xrightarrow{r_2} & \text{PHET} & \xrightarrow{r_4} & \text{ETBE} \\
[B] & & [D] & & [F] \\
\Big\downarrow r_1 & & \Big\downarrow r_3 & & \\
\text{CHMK} & & \text{CHET} & & \\
[C] & & [E] & &
\end{array}
\qquad (2)
$$

The rate equations developed by Mathew (2001) are given below and the corresponding kinetic parameters are listed in Table 2.

$$
R_B = \frac{1}{3} \frac{wk'_1 K_{BY} A_1^* B_1}{(1 + K_{BY} B_1 + K_{DY} D_1)} + \frac{wk'_2 K_{BX} A_1^* B_1}{(1 + K_{BX} B_1 + K_{DX} D_1)} \qquad (3)
$$

$$
R_C = \frac{1}{3} \frac{wk'_1 K_{BY} A_1^* B_1}{(1 + K_{BY} B_1 + K_{DY} D_1)} \qquad (4)
$$

$$
R_D = \frac{1}{3} \frac{wk'_3 K_{BY} A_1^* B_1}{(1 + K_{BY} B_1 + K_{DY} D_1)} + \frac{w A_1^* (k'_2 K_{BX} B_1 - k'_4 K_{DX} D_1)}{(1 + K_{BX} B_1 + K_{DX} D_1)} \qquad (5)
$$

$$
R_E = \frac{1}{3} \frac{wk'_3 K_{BY} A_1^* B_1}{(1 + K_{BY} B_1 + K_{DY} D_1)} \qquad (6)
$$

TABLE 2 Rate Parameters and Temperature Dependency of Acetophenone Hydrogenation (Mathew, 2001)

Rate Constants at 398 K	Activation Energy (kJ/mol)	Adsorption Constants at 398 K	Heat of Adsorption (kJ/mol)
$k'_1 = 2.42 \times 10^{-4}$	53	$K_{BX} = 1.12$	-10.4
$k'_2 = 1.05 \times 10^{-4}$	52	$K'_{BY} = 3.04$	-23.16
$k'_3 = 1.26 \times 10^{-4}$	57	$K_{DX} = 7.8$	-15.77
$k'_4 = 1.48 \times 10^{-4}$	55	$K'_{DY} = 0.40$	-11.43

Reactions are described by Eq. (2).

$$R_F = \frac{1}{3} \frac{w k'_4 K_{DX} A_1^* D_1}{(1 + K_{BX} B_1 + K_{DX} D_1)} \tag{7}$$

Here the subscripts B, C, D, E, and F stand for different reactants as indicated in Eq. (2). w is the weight of catalyst per unit volume (kg/m^3), k'_1–k'_4 are kinetic constants (m^3/kg) (m^3/kmol s), and K_{BX}, K_{BY}, K_{DX}, and K_{DY} are equilibrium adsorption constants (m^3/kmol).

A trickle bed reactor model was developed following the approach discussed in Chapter 3. It is important to note here that reactor hydrodynamics was not modeled and the gas and liquid phases were assumed to follow plug flow behavior. For the gas phase reactant, gas-liquid, liquid-solid, and intraparticle mass transfer limitations were considered. Other assumptions were (1) negligible external as well as intraparticle mass transfer effects for liquid phase components, (2) negligible contribution by stagnant liquid pockets, (3) no radial concentration and temperature gradients within the reactor (isothermal conditions), (4) no heat conduction in the axial direction, (5) the liquid phase reactants are non-volatile but the solvent has significant vapor pressure, and (6) the gas phase behaves ideally. Based on these assumptions, the total reaction rate for a partially wetted catalyst particle was represented as a sum of the contributions due to mass transfer of reactants to the liquid-covered and gas-covered outer surfaces of catalyst as discussed in Chapter 3.

Under conditions of significant intraparticle gradients for the gas phase reactant (H$_2$) and when the liquid phase reactant was in excess, the overall rate of hydrogenation can be expressed as follows:

$$R_A = \eta w \frac{A_g}{H_A} \left[\frac{k_1 B_1 + k_3 D_1}{(1 + K_B B_1 + K_D D_1)} + \frac{k_2 B_1 + k_4 D_1}{(1 + K_C B_1 + K_E D_1)} \right] \tag{8}$$

where, η is effectiveness factor defined as,

$$\eta = \frac{1}{\phi} \left(\coth 3\phi - \frac{1}{3\phi} \right) \tag{9}$$

and ϕ, the overall Thiele modulus is given as:

$$\phi = \frac{R}{3}\left[\frac{\rho_p}{D_e}\left(\left[\frac{k_1 B_1 + k_3 D_1}{(1 + K_B B_1 + K_D D_1)} + \frac{k_2 B_1 + k_4 D_1}{(1 + K_C B_1 + K_E D_1)}\right]\right)\right]^{1/2} \quad (10)$$

where ρ_p is the particle density, R is the radius of the particle, and D_e is the effective diffusivity. The trickle bed reactor model was then used to predict the rate of hydrogenation and conversion of ACPH.

The effect of liquid velocity on the rate of hydrogenation is shown in Fig. 3 for temperatures of 373, 398, and 423 K. The rate of hydrogenation was found to decrease marginally with increase in liquid velocity. With increase in liquid velocity, one expects increase in the wetted fraction of the catalyst as well as increase in the gas–liquid and liquid–solid mass transfer coefficients. The variation of gas–liquid mass transfer and wetting efficiency for the considered liquid velocity range were estimated using the correlations suggested by Fukushima and Kusaka (1977) and Al-Dahhan and Dudukovic (1995), respectively. At lower liquid velocities, catalyst particles are partially wetted (wetted fraction = 0.47 for a liquid velocity of 5.4×10^{-4} m/s) and under these conditions, the rate increases due to direct transfer of gas phase reactant to the catalyst surface (already wetted internally due to capillary forces). Therefore, with increase in liquid velocity, increased wetted fraction is expected to reduce the

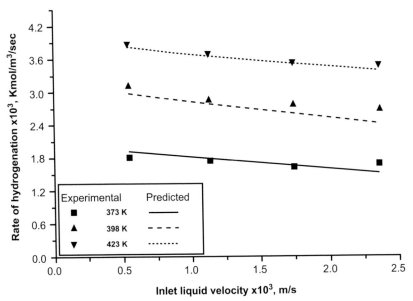

FIGURE 3 Effect of liquid velocity on rate of hydrogenation (from Mathew, 2001). Reaction conditions: reactor pressure: 52 atm, temperature: 398 K, inlet ACPH concentration: 1.27 kmol/m^3, gas flow rate: 30 Nl/h.

rate of reaction. It was found that, under the range of reaction conditions studied by Mathew (2001), the major product was PHET along with considerable amount of CHET. At higher liquid velocities, when the conversions of ACPH were low (~30%), PHET selectivity was high (~78–80%) and at lower liquid velocities (when conversions were high), the selectivity toward CHET increased (~9% at a conversion of 30% and 16% at a conversion of 62%; see Fig. 4).

The influence of gas flow rate for three operating pressures (52, 35, and 20 atm) is shown in Fig. 5. There was only a marginal increase in the conversion as well as rate of hydrogenation with increase in gas velocity at lower pressures. Since the reaction under consideration was limited by gas–liquid mass transfer, one would have expected a stronger effect of gas velocity on the rate of hydrogenation. Estimation of gas–liquid mass transfer coefficient ($k_L a_B$) and the wetting efficiency using the correlations discussed in Chapter 2 may provide better insight. The correlation of Fukushima and Kusaka (1977) was used to estimate mass transfer coefficient and it can be seen that on increasing the gas velocity from 0.035 to 0.23 m/s, value of $k_L a_B$ increases from 1.88×10^{-2} to $2.48 \times 10^{-2}\,\mathrm{s}^{-1}$. However, with further increase in the gas velocity, the increase in the value of $k_L a_B$ is not significant. Similar trends were observed in the variation of conversion with gas velocity shown in Fig. 5. Influence of operating pressure on the reactor performance is illustrated in Figs. 6 and 7. The rate of hydrogenation was

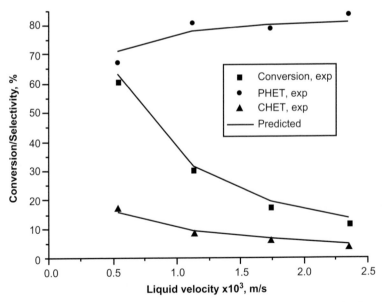

FIGURE 4   Effect of liquid velocity on conversion and selectivity (from Mathew, 2001). Reaction conditions: temperature: 398 K, P_{H2}: 52 atm, substrate concentration: 1.27 kmol/m^3, gas flow rate: 30 Nl/h.

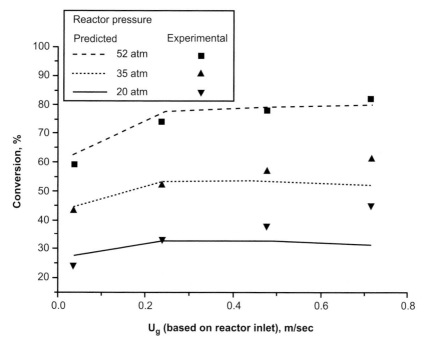

FIGURE 5   Effect of gas velocity on ACPH conversion (from Mathew, 2001). Reaction conditions: U_L: 5.4×10^{-4} m/s, substrate concentration: 1.27 kmol/m^3, wall temperature: 398 K.

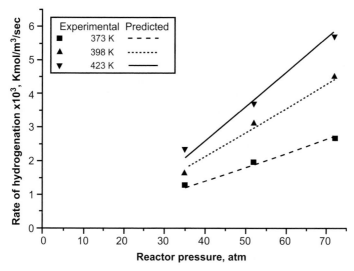

FIGURE 6   Effect of operating pressure on rate of hydrogenation (from Mathew, 2001). Reaction conditions: U_L: 5.4×10^{-4} m/s, inlet ACPH concentration: 1.27 kmol/m^3, wall temperature: 398 K.

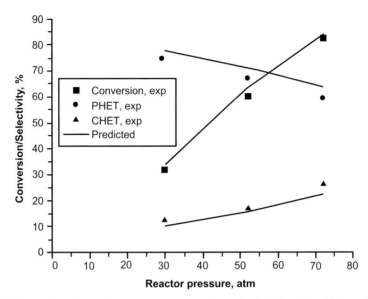

FIGURE 7 Effect of operating pressure on conversion and selectivity (from Mathew, 2001). Reaction conditions: temperature: 398 K, P_{H2}: 52 atm, substrate concentration: 1.27 kmol/m³, gas flow rate: 30 Nl/h.

found to increase with increase in hydrogen pressure (see Fig. 6) with almost a linear dependence. The formation of CHET was higher at higher pressures (26% CHET at a pressure of 72 atm [temperature of 398 K and liquid velocity of 5.4×10^{-4} m/s] whereas at 35 atm it was only ~10%). With increase in temperature, corresponding increase in the formation of CHET was marginal (~11% CHET at 373 K and ~19% at 423 K at a reactor pressure of 52 atm and liquid velocity of 5.4×10^{-4} m/s).

These results show that the partial wetting model captures observed trends of hydrogenation of ACPH adequately in addition to the external and intra-particle diffusion effects. The model correctly captures the key trends observed with respect to operating parameters such as liquid velocity, gas velocity, operating pressure, and temperature.

Hydrogenation of 2,4-Dinitrotoluene (DNT): Complex Exothermic Reaction

Hydrogenation of DNT to toluenediamine (TDA) is an important industrial process in the manufacture of intermediates for polyurethane intermediates and is an example of a complex consecutive reaction with high exothermicity. The reaction scheme for the DNT hydrogenation can be represented as shown below:

$$(11)$$

where, 2,4-DNT: 2,4-dinitrotoluene reaction; 2-A-4-NT: 2-amino-4-nitro-toluene; 4-A-2-NT: 4-amino-2-nitrotoluene, and TDA: 2,4-toluenediamine.

The rate analysis and the reaction engineering model for such a system can be developed using the methodology discussed in Chapter 3 (Non-isothermal Trickle Bed Reactor Model: Complex Reactions). Rajashekharam, Jaganathan, and Chaudhari (1998) have considered the above reaction scheme and presented a detailed reaction engineering model for a trickle bed reactor. The model equations are not repeated here to avoid duplication. Following the methodology discussed in Chapter 3, the intrinsic reaction kinetics was first determined independently. The kinetic model was then used to predict the performance of trickle bed reactor. The model was verified by comparing the predictions with the experimental data obtained from a laboratory trickle bed reactor (see Rajashekharam et al., 1998 for more details). Some of their results on DNT hydrogenation are reproduced here in Fig. 8. This example illustrates a case involving rapid, exothermic reaction with complex multistep reactions and high exothermicity. The other effects like partial wetting and non-isothermal effects leading to significant temperature gradients and hysteresis behavior are also observed and captured well by the proposed model. The reaction engineering model provides better understanding of several complexities as a result of interactions between high exothermicity, mass transfer, and reaction kinetics with hydrodynamics and wetting.

Such reaction engineering models, which are experimentally verified on the laboratory scale, can then be used in association with the scale-up methodologies and wherever necessary with the detailed hydrodynamic models (computational fluid dynamics (CFD)) for understanding and improving performance of larger-scale reactors.

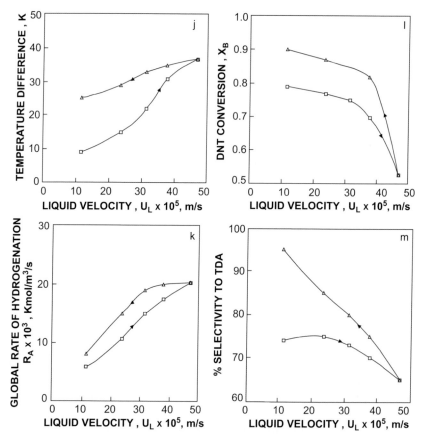

FIGURE 8 Experimental observation of multiplicity of conversion, global rate, temperature rise, and selectivity in a trickle bed reactor (Rajashekharam et al., 1998). Initial conditions: $C_L = 0.5$ kmol/m³; H$_2$ pressure: 3 MPa; $U_G = 4.23 \times 10^{-3}$ m/s; temperature = 328 K.

Hydroprocessing Reactions

Hydroprocessing of petroleum feedstock is an important process for production of fuels, upgradation of synthetic lubricant oils, and coal liquefaction process. Hydroprocessing generally referred to a group of processes used in petroleum refineries, i.e., HDS, hydrodemetallization (HDM), HDN, HDO, which are almost exclusively carried out in trickle bed reactors. These processes are required to control the environmental norms (controlling emission of sulfur and nitrogen), avoiding the poisoning of catalyst used for further refining of fuel and improving stability, color, and odor. Further refining of fuel involves hydrogenation (HYD) and hydrodearomatization (HDA) which improves stability and quality of the product. Aromatics, primarily responsible for quality of fuel (octane number, solubility, color, etc.), are required to be reduced to improve the exhaust gases emission. In petroleum fractions, compounds like sulfur, nitrogen, oxygen, heavy metal, etc. are present in a variety of forms. For example, in diesel oil, sulfur is present in the form of heterocyclic compounds such as benzothiophenes, dibenzothiophenes, and benzonaphthothiophenes. Nitrogen is present in the condensed heterocyclic form which requires high pressure and temperature for reduction. In recent years, awareness of decreasing aromatic content in the fuel is increasing due to adverse effects of exhaust emission. Therefore, designing and operation of trickle bed hydro-processing reactors to achieve stringent product quality have remained major challenges in the last few decades.

Design of hydroprocessing trickle bed reactors is a complex task due to two reasons: (i) multiparametric system with complex hydrodynamics and (ii) lack of data on intrinsic kinetics of individual reactions and deactivation steps. These industrial hydroprocessing reactors are rather large in size. Therefore, heat transfer efficiency, likelihood of hotspots, and liquid maldistribution are the major concerns while designing and scaling-up of these reactors. The temper-ature variation and formation of hotspots will adversely impact sulfur contents in the product and therefore the quality of product. In such cases, reaction engi-neering models used for the examples discussed in previous section may not be adequate. It is often essential to use these models along with detailed CFD-based models. Recent advances in the CFD modeling of trickle bed reactor (for example, Gunjal, Kashid, Ranade, & Chaudhari, 2005; Jiang, Khadilkar, Al-Dahhan, & Dudukovic, 2001) can be used to reduce the efforts in experiments and empiricism in design and scale-up of the trickle bed reactors. CFD models of trickle bed reactors are discussed in the Chapter 4 and if formulated correctly, the models can be used to facilitate design of such large-scale trickle bed reactors. Such models are capable of accounting for several complex behaviors observed in trickle bed reactors such as liquid maldistribution, liquid bypassing, and local segregation of liquid. Recognizing the potential of CFD models in reducing the associated uncertainty with trickle bed hydrodynamics and variation in reactor hardware, illustration of their application for simulating hydroprocessing reactors will be helpful. In this section, the CFD model and results presented by Gunjal and Ranade (2007) are briefly discussed as an example. The purpose is to present the methodology rather than the specific design guidelines for hydro-processing reactors. The methodology can be used with some modifications in the underlying assumptions (if necessary) to simulate industrial hydro-processing reactors.

Various noble metals (platinum, palladium, ruthenium, rhodium) and non-noble metals (cobalt, nickel, molybdenum, and combination of these metals) have been used for hydroprocessing. Selection of these catalysts is based on the composition of feedstock and desired products. Hydroprocessing catalysts are prepared generally with γ-Al_2O_3 support and suitable pore size needs to be maintained to provide the passage for the larger molecular compounds to the active sites of the catalyst. Hydroprocessing is typically carried out in the range of temperature 400–750 K with pressure being 20–80 MPa. In this example, the feedstock composition and reactor configuration similar to that used in the study of Chowdhury, Pedernera, and Reimert (2002) were considered.

Based on the studies of Bhaskar, Valavarasu, Sairam, Balaraman, and Balu (2004) and Chowdhury et al. (2002), two major reactions viz., HDS and HDA were considered in the model and other minor reactions were neglected. Following assumptions were made (see Gunjal and Ranade 2007 for more details):

1. Uniform reactor pressure (pressure drop is insignificant compared to the operating pressure)
2. Trickle bed reactor is operated isothermally (heat of reaction is moderate and heat transfer limitations absent)
3. Gas phase consists of hydrogen and product H_2S and ideal gas law is applicable
4. Liquid phase reactants are non-volatile (negligible vapor pressure)
5. Gas–liquid mass transfer is the limiting resistance with negligible liquid–solid and intraparticle mass transfer limitations
6. The catalyst particles are completely wetted (This assumption may not be valid for larger-sized reactor. The expressions of effective reaction rates used here account for the partial wetting phenomenon)

The following reactions were considered:

$$Mono - Ar + 3H_2 \Leftrightarrow Naphthenes \qquad (12)$$

$$Di - Ar + 2H_2 \Leftrightarrow Mono - Ar \qquad (13)$$

$$Ar - S + 2H_2 \rightarrow Ar + H_2S \qquad (14)$$

$$Tri - Ar + H_2 \Leftrightarrow Di - Ar \qquad (15)$$

The CFD model was developed to get better understanding of the hydrodynamic effects (gas–liquid interaction, bypassing, etc.) and reactor hardware (bed porosity, reactor geometry, etc.). Details of CFD model equations are described in Chapter 4. The predicted flow field was used to simulate performance of trickle bed reactors at different scales.

The following rate expressions, for this case given by Chowdhury et al., 2002, were used:

$$k_{ij}^E = k_{ij,0}e^{-(E_{ij}/R)(1/T-1/T_0)} \qquad (16)$$

$$r_{Poly} = -k_{Poly}C_{Poly} + k_{\_Poly}C_{Di} \qquad (17)$$

$$r_{HDS} = -\frac{kc_{L,H_2}^{0.56}c_{L,S}^{1.6}}{1 + K_{Ad}c_{L,H_2S}} \qquad (18)$$

$$r_{Mono} = -k_{Mono}C_{Mono} + k_{\_Mono}C_{Naph} \qquad (19)$$

$$r_{Di} = -k_{Di}C_{Di} + k_{\_Di}C_{Mono} \qquad (20)$$

TABLE 3 Kinetic Parameters of Hydroprocessing Reactions (from Chowdhury et al., 2002)

Kinetic Constants	Values
K_{Ad} (m³/kmol)	50 000
K (dimensionless)	$2.5 \times 10^{12} \exp(-19\,384/T)$
k^*_{mono} (m³/kg s)	$6.04 \times 10^2 \exp(-12\,414/T)$
k^*_{Di} (m³/kg s)	$8.5 \times 10^2 \exp(-12\,140/T)$
k^*_{poly} (m³/kg s)	$2.66 \times 10^5 \exp(-15\,170/T)$

Kinetic expressions are described by Eqs. (16)–(20).

---

where, r is the rate of reaction in kmol/kg s, C is the concentration of reactant in kmol/m³, and k is the rate constant. Kinetic parameters of above reactions used in the simulations are listed in Table 3.

Mass balance equation for any species, i was written as:

$$\nabla \cdot (\varepsilon_k \rho_k U_k C_{k,i}) = -(\varepsilon_k \rho_k D_{i,m} \nabla C_{k,i}) + \varepsilon_k \rho_k \Gamma_{i,k} \tag{21}$$

where, $C_{i,k}$ is the concentration of i-th species in k-th phase (gas or liquid) and ρ_k and ε_k are the density and volume fraction of the k-th phase, respectively. S_{ik} is the source for species i in phase k.

Volume-averaged properties of fluid were used for calculating fluxes across the control cell. The source term for species i in gas phase was written as:

$$\Gamma_i = -k_{GLi}a_{GL}\left[\frac{C_{Gi}}{H_i} - C_{Li}\right] \tag{22}$$

The corresponding source term for species i in liquid phase was written as,

$$\Gamma_i = k_{GLi}a_{GL}\left[\frac{C_{Gi}}{H_i} - C_{Li}\right] + \rho_B \eta \sum_{j=1}^{j=nr} r_{ij} \tag{23}$$

where, r is the rate of reaction in kmol/kg s, nr is the number of reactions, ρ_B is the catalyst bulk density in kg/m³, and η is the wetting efficiency.

The source term for non-volatile species in liquid phase, because of negligible liquid–solid mass transfer resistance and fast intraparticle diffusion (catalyst effectiveness factor $= 1$), becomes,

$$\Gamma_i = \rho_B \eta \sum_{j=1}^{j=nr} r_{ij} \tag{24}$$

Two configurations of hydroprocessing reactors were considered to demonstrate design issues of two different scales of reactors. Detailed dimensions considered for two scales of reactors are shown in Fig. 9. Diameter of commercial scale reactor was 100 times larger than the laboratory-scale reactor. The real-life design and scaling-up process is usually iterative and complex as discussed in Chapter 5. The purpose of this brief discussion here is to illustrate capabilities of CFD models in handling various issues related to the scaling-up.

One of the key differences in the laboratory and the larger commercial-scale reactors is radial variation of porosity within the bed. For low values of ratio of bed diameter to the particle diameter, the porosity variation near the wall is quite high (see Bey & Eigenberger, 1997; Mueller, 1991; Chapter 4). This porosity variation leads to a significant difference in the flow field at two scales. For example, ratio of maximum interstitial velocities to the superficial velocity for laboratory-scale reactor is quite high (20−25) as compared to the commercial reactor (~1.6). Different aspect ratios at two scales also result in significant differences in liquid holdup at two scales. The developed CFD model was first used to simulate hydrodynamics. The predicted pressure drop and liquid holdup were compared with those estimated from different correlations. After establishing an adequate agreement, the CFD model was used to investigate differences in reactor performance at two scales of operation.

Hydroprocessing reactor performance was simulated by combining CFD model and reaction engineering models. Several numerical experiments were carried out to obtain better understanding of influence of various parameters

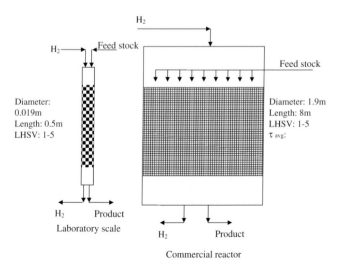

FIGURE 9   Dimensions of the laboratory and commercial-scale reactors considered by Gunjal and Ranade (2007).

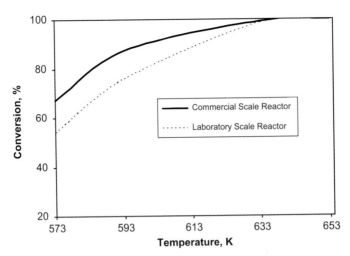

FIGURE 10 Conversion of aromatics sulfur compound at different temperatures (from Gunjal and Ranade, 2007).

and reactor scale on reactor performance. Sample of these results is included here. Performance of the commercial reactor (conversion) is better than that of the laboratory reactor (see Fig. 10). As the reactor operating temperature increases, the difference between the commercial and laboratory-scale reactor reduces (see Fig. 11). The predicted outlet sulfur concentration in the commercial reactor is much lower than that in the laboratory reactor

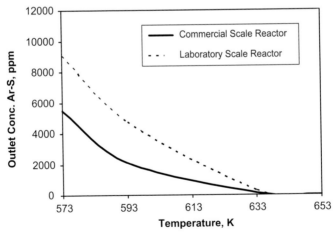

FIGURE 11 Outlet sulfur composition at different temperatures (from Gunjal and Ranade, 2007). $P_{lab \ and \ com} = 4$ and 4.4 MPa, $LHSV_{lab \ and \ com} = 2.0$ and $3.0 \ h^{-1}$, $(Q_G/Q_L)_{lab \ and \ com} = 200$ and $300 \ m^3/m^3$, $y_{H2S} = 1.4\%$.

especially at low temperature as shown in Fig. 11. The conversion of poly-aromatic and total aromatic compounds is almost similar in the laboratory and commercial reactors (Fig. 12). The predicted conversion of the aromatic sulfur compound decreases monotonically with liquid hourly space velocity (LHSV) in the laboratory-scale reactor. However, in the commercial-scale reactor, conversion was found to remain unchanged after certain increase in liquid flow rate (LHSV = 3 h^{-1}) (see Fig. 13). This may be attributed to rate inhibition due to increase in H$_2$S concentration at higher conversion in the commercial-scale reactor. Numerical experiments carried out with the help of the CFD model indicate that influence of gas concentration on conversion of aromatic compounds is rather negligible. The model predictions indicated that the performance of the commercial-scale reactor is better than the laboratory-scale reactor. This effect is especially dominant for conversion of the aromatic sulfur compound. Similar trends were also observed in recent studies of Rodrýguez and Ancheyta (2004). Possible reasons for such improved performance of a larger reactor could be, (a) significantly higher liquid holdup in the larger reactor (~0.2 compared to less than 0.1 in the laboratory reactor) and (b) enhanced gas–liquid mass transfer performance of the larger reactor. The CFD models provide a platform to interpret the laboratory-scale, pilot-scale, and commercial-scale data and to gain better insight. The understanding gained through such modeling exercise facilitates reliable scale-up and better reactor engineering of trickle beds when combined appropriately with accumulated experience [for appropriate selection of distributors, catalyst particle size/shape, and other reactor internals] and detailed rate analysis.

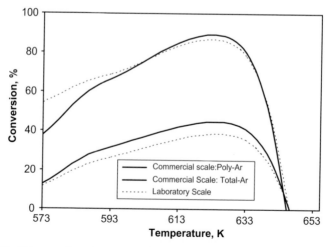

FIGURE 12 Polyaromatic and monoaromatic conversion at different temperatures (from Gunjal and Ranade, 2007).

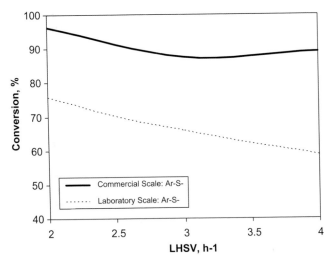

FIGURE 13 Conversion of aromatic sulfur compound at different LHSVs (from Gunjal and Ranade, 2007). $P_{lab\ and\ com} = 4$ and 4.4 MPa, $LHSV_{lab\ and\ com} = 2.0$ and 3.0 h^{-1}, $(Q_G/Q_L)_{lab\ and\ com} = 200$ and 300 m^3/m^3, $y_{H2S} = 1.4\%$.

Oxidation Reactions

Oxidation (liquid phase) reactions with solid catalysts have important applications using trickle bed reactors, some examples of which are SO_2 oxidation to SO_3 for sulfuric acid manufacture and SO_2 removal as a pollutant (Goto & Smith, 1978), wastewater treatment to remove traces of organic pollutants, and oxidation of aqueous organic pollutants from industrial wastes (pollution control). Several studies in oxidation using model reactions like oxidation of aqueous formic acid (Goto & Smith, 1975) and acetic acid (Levec & Smith, 1976) have been reported to illustrate modeling of trickle bed reactors. The most distinguishing feature of these processes is very low liquid substrate concentrations required to be converted completely for environmental regulations. This coupled with sparingly soluble oxygen makes the system as gas and liquid reactant limiting similar to HDS processes. Due to larger catalyst particles being used in trickle beds, the problem also needs to consider multicomponent diffusion with reaction in porous catalysts. The challenge is to optimize the reactor for high conversion in spite of lower intrinsic rates as a result of very low substrate concentrations. However, the oxidation reactions are generally fast reactions compared to many other reactions. To control gas phase oxygen concentration to avoid explosion limits is also a challenge. Oxidations are also highly exothermic reactions, often limited by external as well as intraparticle mass transfer. An excellent review by Goto, Levec, and Smith (1977) discusses various aspects of carrying out oxidation processes in trickle bed reactors. Such complex process considerations require detailed trickle bed reactor models as discussed in Chapters 3 (reaction engineering) and 4 (CFD) for reliable prediction of the reactor performance. Apart from these regular oxidation processes, trickle bed reactors are also finding large applications in catalytic wet air oxidations (CWAOs) for wastewater treatment.

Wastewater treatment uses CWAO for treating organic pollutants. In this process, these pollutants are oxidized by dissolved oxygen in the presence of a solid catalyst. Recent developments of stable catalysts for wet air oxidation have renewed interest in process design and reactor scale-up for these applications (Eftaxias, Larachi, & Stüber, 2003). Trickle bed reactor offers significant advantages (for example, less susceptibility to loss of catalyst activity because of polymerization) over other reactor types for carrying out wet air oxidation process (Pintar & Levec, 1994). In general, the oxidation of organic pollutants follows complex reaction pathways. Therefore reduction of chemical oxygen demand (COD) critically depends on the distribution of intermediates, by-products, and end-products and may be different for the same pollutant conversion. In order to use the heat released by the exothermic oxidation reactions and thereby reduce the net energy consumption of the wet air oxidation process, these reactors are usually operated adiabatically (Debellefontaine & Foussard, 2000). In order to illustrate the approach and methodology presented in this book, the work of Eftaxias et al. (2003) on wet air oxidation of phenol is discussed here briefly.

The reaction scheme considered by Eftaxias et al. (2003) is,

$$C_6H_5OH + 0.5O_2 \rightarrow C_6H_4(OH)_2 \tag{25}$$
Phenol   dihydric Phenol

$$C_6H_4(OH)_2 + 0.5O_2 \xrightarrow{r_2} C_6H_4O_2 + H_2O \tag{26}$$
dihydric phenols   benzoquinones

$$C_6H_4O_2 + 2.5O_2 + H_2O \xrightarrow{r_3} C_4H_4O_4 + C_2H_2O_4 \tag{27}$$
benzoquinones   maleic acid oxalic acid

$$C_6H_4O_2 + 5O_2 \xrightarrow{r_4} CH_2O_2 + C_2H_2O_4 + 3CO_2 \tag{28}$$
benzoquinones formic acid oxalic acid

$$C_4H_4O_4 + O_2 \xrightarrow{r_5} C_3H_4O_4 + CO_2 \tag{29}$$
maleic acid   malonic acid

$$C_3H_4O_4 + O_2 \xrightarrow{r_6} C_2H_4O_2 + CO_2 \tag{30}$$
acetic acid

$$C_2H_2O_4 + 0.5O_2 \xrightarrow{r_7} 2CO_2 + H_2O \tag{31}$$
oxalic acid

$$CH_2O_2 + 0.5O_2 \xrightarrow{r_8} CO_2 + H_2O \tag{32}$$
formic acid

The kinetics of these reactions was obtained (Eftaxias et al., 2001) as the first step of reactor design and scale-up exercise. Various hydrodynamic and transport coefficients were then estimated using the correlations similar to those discussed in Chapter 2. In order to quantify the influence of intraparticle diffusion, a pellet-scale model was developed. The model accounted for the effect of partial wetting without the need for assuming limiting reactants. The model was flexible enough to accommodate systems that lie between the asymptotic cases of mainly gas-limited or liquid-limited reactions. Particle effectiveness factors were estimated based on this pellet-scale model.

These estimated effectiveness factors were then used to develop a reactor-scale model. Balance equations for the gas phase, dynamic liquid, and static liquid were developed by making appropriate assumptions about the mixing in individual phases. Most importantly, the energy balance equation was developed by accounting for likely evaporation of water as:

$$(u_l\rho_l C_{pl} + u_g\rho_g C_{pg})\frac{dT}{dz} + \frac{\varphi}{A}\Delta H^v - \sum \rho_b r_i^{op}(\Delta H_i) \qquad (33)$$

where ΔH_v is the heat of vaporization, ΔH_j is the heat of formation for j-th compound (J/mol), and A is the reactor cross-sectional area (m^2).

The evaporation rate was determined by assuming that gas phase is always saturated with water. The model equations were solved with appropriate boundary conditions and other sub-models (such as heat exchanger model and models for estimation of physical properties and transport coefficients). The comparison of the model predictions and the experimental data obtained on the laboratory-scale reactor is shown in Fig. 14. It can be seen that predicted results show good agreement with the experimental data.

The model was then used to understand the influence of mass transfer and hydrodynamic parameters on phenol conversion. Eftaxias et al. (2001) used the parameter γ, proposed by Khadilkar et al. (1996), to identify whether the reaction is gas-reactant limited or liquid-reactant limited. The parameter γ is defined as:

$$\gamma = (D_{ph}^e C_{ph})/(\nu D_{o2}^e C_{o2}) \qquad (34)$$

where ν is stoichiometric coefficient of reaction, D^e is effective diffusion coefficient, and C is concentration. The subscripts "ph" and "o2" denote phenol and oxygen, respectively. When $\gamma > 1$, the reaction is gas-reactant limited, conversely, when $\gamma < 1$, it turns to liquid-reactant limited. Typical wet air oxidation operating conditions fall in the transition region between gas-limited and liquid-limited reactants. The results of Eftaxias et al. (2001) indicated that phenol conversion increases for partially wetted conditions as compared to the fully wetted condition. The non-isothermal simulations indicated that the extent of water evaporation can significantly influence the adiabatic temperature rise as well as phenol conversion. Naturally, the influence is more significant for the low liquid velocities. The models allow better understanding

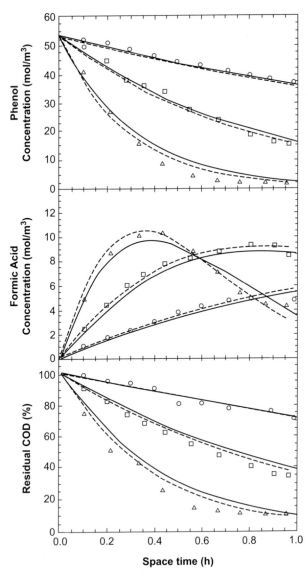

FIGURE 14   Model prediction and experimental data (Fortuny et al. 1999) for the CWAO of phenol in a laboratory-scale trickle bed reactor (TBR) at 0.6 MPa of O_2. (a) Phenol, (b) formic acid, and (c) residual COD. Lines indicate models: —, TBR model; - -, kinetic model (Eftaxias et al., 2001). Points indicate experimental data: (x) 120 °C, (x) 140 °C, (x) 160 °C (from Eftaxias et al., 2003).

of complex influence of inlet concentration of phenol or inlet temperature of liquid (because of opposing effects of heat released by chemical reactions and water evaporation) on reactor performance.

The basic reactor models illustrated in this example can be complemented by CFD models to gain better understanding of wetting, liquid distribution, and heat transfer. The models and methodology discussed in Chapter 4 will be useful for this purpose. Many more examples from our consulting experience or from published literature can be discussed here. However, the approach and methodology are more or less same as discussed in Chapters 2–5 and illustrated in some of the examples discussed here. Therefore, further application examples are not included here. Instead some recent developments on variants of trickle bed reactors are discussed in the following.

RECENT DEVELOPMENTS

Trickle bed reactors have been used in process industries for several decades. Considering the specific requirements of different applications, some variants of trickle bed reactors are used in practice. One of the most successful variants is the monolith reactor. In recent years, miniaturized (micro-channel) trickle beds have been proposed and used for small-scale continuous processes. These two promising variants of classical trickle bed reactors are discussed in this section.

Monolith Reactors

Monolith catalyst concept has been extensively used in automotive industries for reducing exhaust gas emissions. Realizing several potential advantages, an idea of using monolith reactor (catalyst) for carrying out gas–liquid reactions was suggested by Satterfield and Ozel (1977). Monolithic reactors offer advantages such as low pressure drop, high mass transfer rates, and easier scale-up. More importantly, the active catalytic material is coated externally and therefore extent of pore diffusion limitations are significantly less than the conventional trickle bed reactor. This helps in effective utilization of the catalysts. Tedious fabrication process and possible difficulties in uniformly distributing gas and liquid phases within monoliths are some of the disadvantages. Significant work on investigating monolithic reactors has been done (see a review by Roy, Bauer, Al-Dahhan, Lehner, & Turek, 2004; and a book by Cybulski & Moulijn, 2006). Key aspects of monolith reactors are briefly discussed here.

Design and Construction

"Monolith" represents a single block of ceramic or metal containing several straight or corrugated channels. Catalyst is impregnated or coated on the walls of the channel where gas–liquid phases react with each other. Monoliths are made either by extruding single block metal or by arranging sheets to form parallel (straight or corrugated) channels. Flow channels are extended over entire block separated by thin walls. Therefore there is no interaction of fluid with neighboring channels. Structure and shape of the channels can be made up of variety of shapes to manipulate flow, surface area, hydrodynamics, and heat/mass transfer rates. These shapes can be triangular, hexagonal, square, circular, honeycomb, etc., as shown in Fig. 15a. Besides shape and size of the passages, thickness of wall (catalyst volume) can be manipulated to achieve the desired goals. Size of channel is measured in terms of number of cells per square inch. Typical monoliths used in industry have 200–800 channels per square inch (cpsi). Monoliths are characterized by defining voidage (volume of catalyst per unit bed volume), open frontal area (OFA), geometric surface area (GSA), and hydraulic diameter. Performance of the monoliths can be manipulated by appropriate selection of these parameters.

Classically monolith reactors are designed as single block of metal/ceramic of length from 0.5 to 5 m. Gas–liquid phases, flow cocurrently downwards with some part of the gas/liquid phases being recirculated along with make-up liquid as shown in Fig. 15b. In case of exothermic reactions, coolants are passed through certain set of channels (inline cooling) or external loop through heat exchanger (offline cooling). For monolith reactors, initial liquid distribution is important because there is no lateral interaction. It is possible and often desirable to use multiple monoliths with liquid redistributors. For a given volume of catalyst, gas–liquid phases are contacted more efficiently in monoliths than in conventional trickle beds. For example, smaller-sized particles in trickle bed reactors provide higher catalyst surface area, mass transfer rates, and lower diffusion effects at the expense of higher pressure drop. The same effect can be achieved by using monolithic reactors with lower

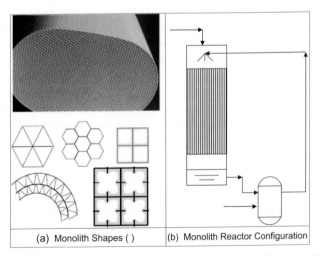

| (a) Monolith Shapes ( ) | (b) Monolith Reactor Configuration |

FIGURE 15 Structure of monolith and reactor configuration. (a) monolith shapes; (b) monolith reactor configuration.

pressure drop. Radial cross-flow liquid segregation and maldistribution observed in trickle bed reactors are less likely to occur in monolithic reactors if the initial distribution is effective. Monoliths also offer better advantages in terms of operability and process control.

Apart from difficulties associated with uniform liquid distribution over large number of channels, long channel length may lead to unstable two-phase flows. This may require stacking of monoliths with smaller length of channels. Stacking and liquid distribution need to be handled carefully while scaling-up of monolithic reactors. Possible implications of cost of monolith/catalyst coating and deactivation of catalyst need to be evaluated. Key aspects of hydrodynamics, reactor engineering, and applications of monolith reactors are discussed in the following.

Fluid Mechanics

Two-phase flow in monolith reactors needs to be considered at two scales, i.e., reactor scale and local/channel scale. Not much information is available on reactor-scale hydrodynamics (complete monolith block) whereas relatively more information is available on channel scale. Most of the studies on channel scale were carried out with a circular channel. Four distinct flow regimes have been observed in two-phase flow in channels (Irandoust & Anderson, 1988) irrespective of channel shape, (circular, square, honeycomb): (i) bubbly flow, (ii) slug flow, (iii) Taylor flow, and (iv) churn turbulent flow (see Fig. 16).

Bubbly flow regime is observed at low gas and high liquid flow rates where liquid phase is continuous and gas is in the form of dispersed bubbles. At moderate gas flow rate and high liquid flow rates, individual bubbles collapse and form single bubbles whose diameter is restricted by channel diameter.

Further increase in gas velocity results in increased length of the bubble and this flow regime is referred as *Taylor flow*. Taylor flow with gas entrainment at the bottom of the bubbles is referred as *slug flow*. In both of these flow regimes, gas flow in the form of large bubbles is separated by liquid slugs. Taylor flow has gained more attention due to well-defined bullet shape gas bubbles with internally circulating laminar flow inside the liquid slug (see Fig. 17). Internal circulation improves mass transfer. Separated slugs reduce back mixing. Thin liquid film is formed between moving gas bubble and wall of the channel. This film gets replenished by gas and liquid phase reactants when it comes in contact with a moving bubble and a liquid slug respectively.

Slug length and film thickness are two key parameters which can be manipulated to achieve the desired performance. Within operating regime of Taylor flow, *slug length* is a function of gas−liquid flow rate and diameter of the channel. Following relation may be used,

$$L_S = \psi d_h \tag{35}$$

where d_h is hydraulic diameter of the monolith channel and ψ values vary in the range of 2−5 depending on the system. Heiszwolf (2001) and Kreutzer

FIGURE 16 Flow regimes: (a) bubbly flow; (b) slug flow; (c) Taylor flow; (d) churn turbulent flow, reproduced from Irandoust and Anderson (1988).

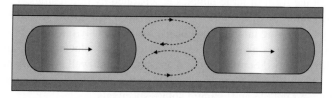

FIGURE 17 Gas—liquid flow in the Taylor flow regime.

(2003) suggested following correlation based on experimental data for 200 cpsi monolith for the estimation of ψ:

$$\psi = \frac{\varepsilon_L}{-0.00141 - 1.55\varepsilon_L^2 \ln(\varepsilon_L)} \qquad (36)$$

Laborie, Cabassud, Durand-Bourlier, and Laine (1999) have suggested expression based on gas phase Reynolds number $(Re_G = (\rho_G U_G d_c)/\mu_G)$ and Eotvos number $(Eö = ((\rho_L - \rho_G)d_C^2 g/\sigma))$ to calculate the slug length as,

$$\psi = 3451\left(\frac{1}{Re_G Eö}\right)^{1.2688} \tag{37}$$

Liu, Zhao, Li, and Ji (2005) have suggested a correlation for slug length based on experimental data measured on 1−3 mm circular monolith as:

$$\frac{U_{TP}}{\sqrt{L_S}} = 0.008\, Re_G^{0.72}\, Re_L^{0.19} \tag{38}$$

where U_{TP} is the two-phase velocity.

It should be noted that slug formation is a complex process and is quite sensitive to the inlet configurations. This may be a reason for possible large variation in the values of slug length estimated from the correlations listed above.

Film thickness is a function of gas−liquid properties, especially viscosity and surface tension (expressed in terms of Capillary number, $Ca = U\mu/s$). Based on the lubrication theory, Bretherton (1961) has derived an expression for determining the film thickness for low as well as high capillary numbers as follows:

$$\frac{\delta}{d_C} = 0.66(Ca)^{2/3} \quad \text{for} \quad Ca < 0.001 \tag{39}$$

$$\frac{\delta}{d_C} = \frac{0.66(Ca)^{2/3}}{(1 + 3.33(Ca)^{2/3})} \quad \text{for} \quad Ca > 0.001 \tag{40}$$

where d_C is a channel diameter.

Irandoust and Andersson (1989) have derived a correlation based on their experimental data which covers a range of Capillary number up to 1×10^{-6} to 10).

$$\frac{\delta}{d_C} = 0.18 \times \left\{1 + \exp\left(-3.08(Ca)^{0.54}\right)\right\} \tag{41}$$

The estimated values of film thickness from both these correlations are close by for higher capillary numbers (see Fig. 18). Film characteristics in Taylor flow (film thickness, velocities in film) become important in reaction engineering analysis.

For square channel, Taylor bubble behavior is slightly different from that in cylindrical channel. Film thickness in square channel is relatively larger than cylindrical channel and film cannot be considered as stagnant even for smaller capillary number. Bubble diameter in square channel is slightly lower than the channel diameter ($d_b/d_C = 0.96$) for Ca ~ 0.04 and it can be up to a maximum of 1.2 as it is measured along the diagonal of a square channel. Following

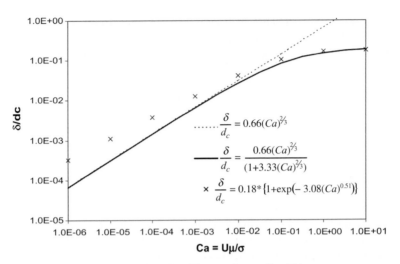

FIGURE 18 Effect of capillary number on film thickness.

correlations can be used for estimation of bubble diameter in square and circular channels.

$$\left[\frac{d_b}{d_C}\right]_{square} = 0.7 + 0.5\, \exp\left(-2.25 Ca^{0.445}\right) \qquad (42)$$

$$\left[\frac{d_b}{d_C}\right]_{circular} = 0.64 + 0.36\, \exp\left(-2.13 Ca^{0.52}\right) \qquad (43)$$

In Taylor flow, bubble moves faster than average velocity of gas–liquid phases. Average gas–liquid velocity can be estimated as,

$$U_{TP} = U_G + U_L \qquad (44)$$

For low capillary numbers, the bubble velocity can be estimated as (Cybulski & Moulijn, 2006):

$$\frac{U_{bubble}}{U_{TP}} \approx 1.29 (3Ca)^{2/3} \qquad (45)$$

Liu et al. (2005) have proposed following general correlation for estimation of the bubble rise velocity:

$$\frac{U_{bubble}}{U_{TP}} = \frac{1}{1 - 0.61 Ca^{0.33}} \qquad (46)$$

Two-phase pressure drop, ΔP_{TP}, in monolith channel is composed of three parts (Satterfield & Ozel, 1977) viz., pressure drop occurring due to frictional pressure drop (ΔP_f), static head of fluid (ΔP_{st}), and due to end effects (entrance or exist losses, ΔP_{end}):

$$\Delta P_{TP} = \Delta P_f + \Delta P_{st} + \Delta P_{end} \qquad (47)$$

It is relatively simple to express entrance or exit losses in terms of orifice equation with two phase gas-liquid flow (Satterfield and Ozel, 1977)

$$\Delta P_{end} = \frac{N(V_{or}^2 - V_T^2)}{2}[\varepsilon_L \rho_L + (1 - \varepsilon_L)\rho_G] \qquad (48)$$

where V_{or} is orifice velocity and V_T is total velocity ($V_L + V_G$).

Total fluid head is composed of head developed by gas and liquid phases. Considering orders of magnitude lower density of gas phase, the static head can be estimated as:

$$\Delta P_{st} = -\varepsilon_L \rho_L g L \qquad (49)$$

A variety of approaches have been used to express the frictional pressure drop of slug flow in a channel including the Lockhart–Martinelli-based approach (which assumes two-phase pressure drop is proportional to single-phase pressure drop, i.e., either gas phase or liquid phase flow). Second approach is based on effective properties of the gas and liquid phases (volume-averaged density and viscosity) to estimate the two-phase pressure drop. In the third approach extension of single-phase flow frictional losses to two-phase flow has been found to be a most suitable way for estimation of the two-phase pressure drop of a slug flow in a channel. The expression takes the following form,

$$\Delta P_f = \varepsilon_L f_{TP} \frac{L}{d_C} \frac{1}{2} \rho_L U_{TP}^2 \qquad (50)$$

Cybulski and Moulijn (2006) have shown that f_{TP} is a function of Reynolds and capillary numbers as:

$$f_{TP} = \frac{16}{Re}\left[1 + 0.17\frac{d_C}{L_S}\left(\frac{Re}{Ca}\right)^{1/3}\right] \qquad (51)$$

Taylor flow characteristics vary considerably with gas–liquid flow rates, diameter of channel, and properties of the liquid (viscosity and density). With increase in gas flow rate, frequency of slugs increases resulting in higher gas–liquid interaction and therefore higher pressure drop. Presence of liquid film increases energy loss due to viscous dissipation which is higher for square channel than circular channel. At higher bubble velocity, bubble cap flattens and flow instability increases at the base of the bubble which results in higher pressure loss in the bubble region. One may observe coalescence of bubbles along the length with increase in gas velocity. Flow behavior for longer slug flow is closer to the single-phase flow through channel because of little interaction between successive gas bubbles. Expressing two-phase pressure drop in a similar fashion as the single-phase flow, has been found to be most suitable:

$$\Delta p_{TP} = \varepsilon_L f_{TP} \frac{L}{d_c} \frac{1}{2} \rho_L U_{TP} \qquad (52)$$

For single-phase flow:

$$f_{SP} = \frac{16}{Re} \qquad (53)$$

Extension to two-phase friction coefficient is given by Heiszwolf et al. (2001),

$$f_{TP} = \frac{C_1}{Re_{TP}} \qquad (54)$$

where $C_1 = 18$, 22, and 24 for 200, 400, and 600 cpsi monoliths (see Fig. 19).

One of the key advantages of monolithic reactors is higher rates of external *mass transfer* achievable with lower power input per unit volume than other multiphase reactors such as bubble columns or stirred tank reactors (Heiszwolf et al., 2001). Mass transfer behavior is rather complex and comprises of combination of series and parallel mass transfer processes. Gas phase reactants transfer from Taylor bubble interface to the liquid phase by two possible mechanisms: (i) gas–liquid mass transfer along the bubble length to the thin film present between a bubble and the channel wall $(k_{GL})_f$ and (ii) gas–liquid mass transfer through bubble caps to the liquid slug $(k_{GL})_{SL}$. Since velocities in the film and bulk of the slug are considerably different, liquid–solid mass transfer rates are also significantly different. The liquid–solid mass transfer rates can also be considered as two separate processes: (i) mass transfer from film to solid surface $(k_{LS})_f$ and (ii) mass transfer from bulk of the slug to the wall $(k_{LS})_{SL}$. For thinner films, the direct mass transfer from Taylor bubble to the walls of the channel can be considered as gas–solid mass transfer $(k_{GS})_f$ instead of

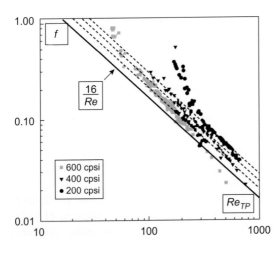

FIGURE 19 Single-phase and two-phase flow pressure drops in a monolith channel (Heiszwolf et al., 2001).

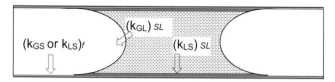

FIGURE 20   Mass transfer in Taylor bubble flow.

considering $(k_{GL})_f$ and $(k_{LS})_f$ (Kreutzer, Du, Heiszwolf, Kapteijn, & Moulijn, 2001). Overall mass transfer mechanisms are schematically shown in Fig. 20.

Bercic and Pintar (1997) developed a correlation for gas–liquid mass transfer in Taylor flow as a function of two-phase velocity, liquid holdup, and unit length as:

$$k_{GL}a_{GL} = \frac{0.111u_{TP}^{1.2}}{(\varepsilon_L(L_{slug} + L_{bubble}))^{0.57}} \frac{D_A}{D_{methane}} \tag{55}$$

Above expression suggests that mass transfer rate is independent of the column diameter. This is supported by the experimental studies of Kreutzer et al. (2001). van Baten and Krishna (2004) have derived a correlation for gas–liquid mass transfer coefficient based on a fundamental approach in which gas–liquid mass transfer in the film was estimated using the penetration theory. Later it was shown (Vandu and Krishna, 2005) that for fast moving bubbles (high velocities), the film is deprived of substrate due to shorter contact time. Following models were proposed for gas–liquid mass transfer in the film and the slug (Vandu and Krishna, 2004).

$$(k_{GL}a_{GL})_f = \frac{4L_f}{dL_{UC}} \begin{cases} 2\sqrt{\dfrac{D}{\pi t_f}} \dfrac{\ln(1/\Delta)}{1-\Delta} & \text{for } Fo < 0.1 \\ 3.41\dfrac{D}{\delta} & \text{for } Fo > 0.1 \end{cases} \tag{56}$$

where L_f is length of film and L_{UC} is length of unit cell.

$$Fo = \frac{D}{t_f\delta^2} \quad \text{with} \quad t_f = \frac{L_f}{U_{bubble}}$$

and

$$\Delta = 0.7857 \exp(-5.212Fo) + 0.1001 \exp(-39.21Fo)$$

$$(k_{GL}a_{GL})_{SL} = \frac{8\sqrt{2}}{\pi L_{UC}}\sqrt{\frac{DU_{bubble}}{d_C}} \tag{57}$$

Liquid–solid mass transfer step comprises of mass transfer from film between bubble and channel wall and mass transfer from slug to the channel wall. For low capillary number, one can assume very small film thickness and liquid–solid mass transfer resistance need not be considered. For such cases, mass transfer can be represented from gas to solid and gas–solid mass transfer can be estimated as (Kreutzer et al., 2001):

$$(k_{GS}a_{GS})_f = \frac{D}{\delta}\frac{4(1-\varepsilon_L)}{d_C} \quad \text{where} \quad a_{GS} = \frac{4(1-\varepsilon_L)}{d_C} \quad (58)$$

For high capillary numbers or non-circular channel, film thickness can affect the liquid–solid mass transfer rates. In liquid slug, mass transfer rate is a function of internal circulation and length of the slug. Internal recirculation of slug causes a significant enhancement of mass transfer. It is often suitable to compare single-phase mass transfer rates and to calculate enhancement in mass transfer due to presence of bubbles. One can establish the relationship between Sherwood number (based on liquid–solid mass transfer coefficient) and Graetz number with an additional parameter for two-phase monolithic flow as slug length (Cybulski & Moulijn, 2006):

$$Sh = \sqrt{[a(\psi)]^2 + \frac{b(\psi)}{Gz}} \quad (59)$$

where,

$$a = 40(1 + 0.28\psi^{-4/3}), \quad b = 90 + 104\psi^{-4/3}, \text{ and } Gz = (L/d)RePr$$

van Baten and Krishna (2005) carried out detailed analysis of liquid–solid mass transfer using the CFD simulations. Their analysis shows that contribution of film is significant for larger values of Graetz number ($Gz > 10^{-04}$):

$$Sh = \frac{\beta}{Gz_{tubr}^{\alpha}} \quad (60)$$

where,

$$\alpha = 0.61Gz_{slug}^{0.025} \quad \text{and} \quad \beta = \frac{0.5}{(Gz_{slug}/\varepsilon_G)^{0.15}}, \quad \text{and} \quad Gz_{slug}^{0.025} = \frac{L_{slug}D_e}{d_C^2 u_b}$$

Kreutzer et al. (2001) have shown that Nusselt or Sherwood number decreases sharply with increase in Graetz number as shown in Fig. 21. Asymptotic values for Nu or Gz are 20 and 1000 for $Gz/Nu \rightarrow \infty$ and $Gz/Nu \rightarrow 0$, respectively. For single-phase flow, these values are order of magnitude lower than (30 and ~1) the two-phase monolithic flow. Thus, even for liquid–solid reactions with mass transfer limitations, monolith reactors with gas sparging are attractive alternative to conventional reactors.

Reactor Engineering of Monolithic Reactors

Reactor models for gas–liquid flow in monolithic reactors are relatively straightforward compared to the trickle bed reactor because of the well-defined

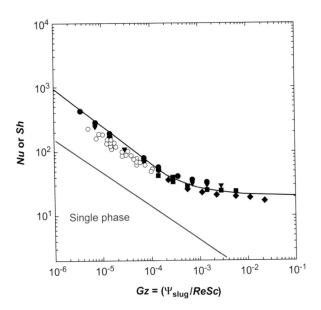

FIGURE 21 Single-phase and two-phase Sherwood or Nusselt number for monolith reactors in Taylor flow regime.

flow pattern inside the channel (Taylor flow or film flow). Various modeling approaches used in the literature are based on analysis of single monolith channel either for film-type flow or Taylor flow. The basic concepts and the model equations are discussed in detail by Albers, Cybulski, Kreutzer, Kapteijn, and Moulijn (2006). Here, mainly Taylor flow reactor engineering model is discussed because its extension to film flow and multichannel monoliths is relatively straightforward. Overall process of mass transfer and reaction is shown in Fig. 22.

Chemical reactions carried out in monolith reactors may be expressed in the following general form:

$$A(g) + \nu B(l) \rightarrow C(l) + D(l) \tag{61}$$

The reaction kinetics may be expressed in terms of Langmuir–Hinshelwood type of rate expression as:

$$\Gamma_A = \nu \Gamma_B = -\frac{k_A k_B C_A C_B}{1 + K_C C_C + K_D C_D} \tag{62}$$

In the bulk region, mass balance of liquid components is simply in terms of mass transfer from bulk of the liquid to the surface of solid catalyst and can be written as:

$$-u_L \frac{\partial C_{BL}}{\partial z} = k_{LS} a_S (C_{BL} - C_{BS}) = \Gamma_B \tag{63}$$

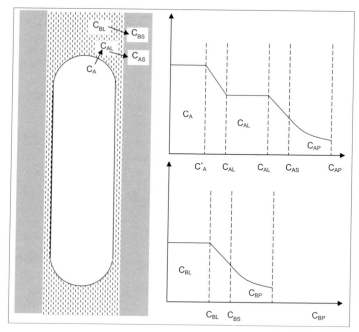

FIGURE 22   Mass transfer in Taylor bubble flow.

Mass balance of gas phase reactant dissolved in the liquid phase is given by,

$$-u_L \frac{\partial C_{AL}}{\partial z} = -k_{GL} a_{GL}(C_A^* - C_{AL}) + k_{LS} a_S (C_{AL} - C_{AS}) = \Gamma_A \qquad (64)$$

Correlations for estimating parameters appearing in the model equations are discussed in the previous section. If the catalyst coat is thinner, then diffusion effects inside the coat can be neglected. If the diffusion effects are not negligible, a detailed model which accounts for diffusion with simultaneous reaction in the porous coat is required. This will allow design of optimum catalyst thickness suitable for monolith reactor. Such approach is explained by Roy et al. (2004). This review by Roy, Bauer, et al. (2004) and Roy, Heibel, et al. (2004) and the chapter by Albers et al. (2006) may be referred for more details of model equations and boundary conditions as well as various different scenarios as per the relative rates of mass transfer and chemical reactions.

Applications

Monolithic multiphase reactors are finding applications in various industrial chemical processes. For example, several hydrogenation reactions can be carried out in monolithic reactors, i.e., hydrogenation of AMS (Heiszwolf et al., 2001; Mazzarino & Baldi, 1987), hydrogenation of glucose, cinnamaldehyde, and benzaldehyde (Heiszwolf et al., 2001), and hydrogenation of nitrobenzoic acid (Hatziantoniou & Andersson, 1984). Monoliths are being used for many oxidation reactions such as aqueous phase oxidation of acetic acid, oxidation of phenol, and oxidation of cyclohexanone (Crezee, Barendregt, Kapteijn, & Moulijn, 2001). Hydrogenation of anthraquinone to the corresponding hydroquinone has been successfully carried out in a monolith reactor with improved performance (Albers et al., 2001). In commercial-scale operation, Akzo Nobel is producing hydrogen peroxide using the monolith reactors (Albers et al., 2001). Many other applications of monolith reactors are described in the monograph by Cybulski and Moulijn (2006) and in the review by Roy, Bauer, et al. (2004). Some of the key issues in limiting the wider applications of multiphase monolith reactors are the lack of suitable and cost-effective distribution system, cost involved in preparation of catalyst, and low radial heat transfer rates. Further developments and innovative modifications in monolith reactors are needed to develop monoliths as a credible alternative to the conventional trickle bed reactors.

Micro-trickle Bed Reactors

Micro-reactors are typically reactors with characteristic length scales ranging from a few microns to less than 1 mm. Such miniaturization provides several advantages in terms of enhanced rates of transport processes (mixing, heat and mass transfer rates) which may be of order of magnitude larger compared to the conventional reactors. These reactors also offer advantages like better control over residence time distribution (RTD), heat and mass transfer, better process flexibility, easier scale-up, and enhanced *process safety*. Micro-reactors are therefore being increasingly explored to implement a variety of applications in chemical process industries. For carrying out gas–liquid–solid reactions, miniaturized trickle bed or micro-trickle bed reactors are used. Key aspects of these reactors are discussed in the following.

Design and Construction

Typical micro-trickle bed reactor is composed of three main parts viz., inlet section, packed bed section, and outlet section. Objective of inlet section is to mix gas–liquid phases and deliver it to each channel uniformly. Middle section contains catalyst in the form of either coated on the walls of the channels or packed in the form of porous particles. Gas–liquid phases are either separated after reaction or fed to another unit for achieving higher residence time and conversion of reactants. Micro-packed bed reactors are aimed to use microsized particles in micro-channels so that process can be scaled-up by numbering up to multiple units in parallel. Two types of micro-reactors can be used for gas–liquid–solid reactions viz., micro-channels with catalyst coated on the wall of the channels and micro-channels filled with micro-sized porous particles. In the first case, catalyst is coated in the form of a film of few microns thick on channel walls. Each channel is similar to the channel of the monolithic reactor. In the second type of reactors, porous particles are packed inside the channel and gas—liquid phases flow over the particles. A typical micro-packed bed reactor made at MIT is shown in Fig. 3 (Chapter 1). Porous particles of micron sizes (typically 25—50 µm) can be placed into micro-channel (400—800 µm) forming the packed column in which gas—liquid—solid reactions can be carried out. In this micro-fabricated packed bed reactor, high surface to volume ratio is achievable to enhance the mass transfer rates.

In the inlet section, gas and liquid phases are separately fed to the distribution system from which they are split into several channels. These channels deliver the gas—liquid phases to the reaction chamber. At the exit of each channel, series of posts are etched to support the catalyst particles. Reaction mixture at the exit of the reactor is either fed to another unit to achieve desired conversion or to the gas—liquid separator.

Similar to the conventional trickle bed reactors, micro-trickle bed reactors may also contain additional hardware components such as thermo-sensors, sampling ports, heating or cooling channels, and gaskets to avoid leakages. Heating and cooling arrangements can be provided either by assigning some of the channels for flowing heat-carrying fluid in it or by jacket composed of flat plate above and below the micro-reactor unit. Fabrication process and material of construction of micro-reactor depend on the properties of fluids used, nature of the reaction (exothermic and endothermic, series, parallel), and the safety aspects associated with the system under consideration. Additional details of design and construction of micro-trickle bed reactors may be obtained from the monograph of Ehrfeld, Hessel, and Löwe (2000) and from Jenson (2001). Key aspects of reactor hydrodynamics, reaction engineering, and applications are briefly discussed in the following.

Fluid Mechanics

Depending upon the gas—liquid feed arrangement and their flow rates, several flow regimes are observed in micro-reactors. Flow characteristics of wall-coated micro-reactor are similar to those observed for monolithic channel flow. Flow regimes like film flow, Taylor flow, bubbly flow, annular flow, ring flow, and churn flow have been observed in such wall-coated micro-reactors (Haverkamp et al., 2006). The film flow and Taylor/bubbly flow regimes are found to be most suitable for micro-reactor operation.

In *falling film-type flow*, liquid film flows along the walls of the micro-reactor under the gravity and gas flows in the core section of the micro-channel. Typical film thickness observed in the micro-reactor is less than 100 µm (in a channel of 300 µm width). Liquid film thickness is a function of gas—liquid properties, flow rates, and the channel geometry. Most of the available correlations are based on film thickness measured for larger-scale channels (3—6 mm hydraulic diameters) and usually overpredict the film thickness in the micro-channel. In micro-reactors, surface forces have larger influence on film hydrodynamics than inertial or gravitational forces. Compared to the conventional trickle bed reactor, mass transfer coefficient ($k_L a$) is higher in film micro-reactor ($\sim 3-8$ s^{-1}, Hessel, Löb, & Löwe, 2005). From operational point of view, falling film micro-reactors require flow rates and liquid distribution to be adjusted in such a way that liquid film should form at the entrance of the channel and be maintained throughout the length of the channel. Surface instabilities, channel overloading, and drying out of film are commonly observed problems in the operation of the film-type micro-reactor.

In *dispersed flow micro-reactor*, separate gas/liquid streams are merged in the mixing section where dispersion occurs and dispersed gas phase causes bubble or Taylor-type flow inside the capillary channel. This phenomenon is similar to the dispersed flow in the monolithic channel; however, in this case channel and bubble sizes are quite small. Liquid phase mixing becomes important in slug flow micro-reactors, since reaction occurs at the wall of the channel. In slug flow, liquid slug is associated with recirculation and thin film separates the gas in the bubble and the catalyst on the wall of the channel. These characteristics of flow not only minimize the axial dispersion but also enhance radial mixing which is required for mass transfer between the phases. Therefore, similar to the monolithic reactors, the Taylor or slug flow is preferred to achieve better performance of the micro-reactors. Pressure drop in slug flow is higher than single-phase liquid flow due to the presence of additional gas phase in the reactor. The pressure drop and liquid holdup can be estimated using the correlations presented for monolith channels (Kreutzer, Kapteijn, Moulijn, Kleijn, & Heiszwolf, 2005). For the Taylor flow, analysis of bubble length variation with gas flow rate is reported by Amador, Gavriilidis, and Angeli (2004) and calculation of two-phase pressure drop and bubble rise velocity in micro-channel is presented by Warnier, De Croon, Rebrov, and Schouten (2010).

In *packed bed micro-reactors*, either array of micro-fabricated structures are incorporated in the channels or channel is filled with porous powder as the catalyst (Hessel et al., 2005). When micro-fabricated structures are incorporated into the micro-channel, then flow regime patterns are similar to those observed in flow through channels, i.e., annual flow, slug flow, and dispersed flow. At low gas flow rates, liquid preferentially flows near the wall with gas as a continuous flow similar to the trickle flow in conventional trickle bed reactors, and gas–liquid interface is comparatively stable. At elevated gas flow rates, considerable fluctuations in gas–liquid interface were observed which can be referred as churn flow (see, for example, Losey, Jackman, Firebaugh, Schmidt, & Jensen, 2002; Wada, Schmidt, & Jensen, 2006; Wilhite, Moreno, & Kim, 2008). For micro-packed bed reactor, two main flow regimes were observed by Losey, Schmidt, and Jensen (2001). The transition from the trickle to the pulse flow regime was also observed albeit at different operating conditions (see Fig. 23). In micro-packed bed reactors, two-phase pressure drop is considerably higher than conventional trickle bed reactors. In this case, gas–liquid phases are intimately contacted with the porous powder in relatively small volume. These micro-trickle

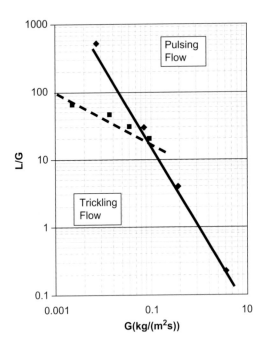

FIGURE 23 Comparison of flow transition from trickle to pulse flow regime in micro-packed bed reactor and conventional trickle bed reactor (Losey et al., 2001). ■ — micro-reactor, cyclohexene/H_2, ♦ — Charpentier and Favior (1975) flow map.

bed reactors offer substantially large interfacial area ($\sim 10\,000-15\,000$ m^2/m^3) and mass transfer coefficient $\sim 1-10$ s^{-1} compared to the conventional trickle bed reactors ($\sim 150-500$ m^2/m^3 and $\sim 0.01-0.05$ s^{-1}, respectively).

Micro-packed bed reactors are recently developed and therefore available information on flow characteristics is rather limited. The brief review and discussion presented in this section may provide useful starting point for designing micro-trickle bed reactors. Some aspects of reaction engineering of the micro-packed bed reactor are discussed in the following.

Reactor Engineering

Reaction models for micro-reactors are based on similar concepts used for trickle bed reactors and monolithic reactors. The first step of reaction engineering analysis is determination of prevailing flow regime based on the operating gas—liquid flow rates and properties of the fluids. If wall-coated micro-reactor is used for the given reaction system, reactor parameters need to be selected in such a way that desired flow pattern is established. If reactor is operated in a film flow regime, accurate estimation of film thickness is crucial. Ensuring stable and uniform film over the entire length of the reactor is also a key issue. The thickness of catalyst wall coat in micro-reactors is usually of few microns. Therefore, usually diffusion effects are not important and plug flow-type reaction engineering models are generally adequate to represent micro-trickle bed reactors. The reactor model equations may take similar form as those of monolithic reactor model (see Section Reactor Engineering of Monolithic Reactors, and references cited therein). Application of such models for micro-reactors is demonstrated by Kiwi-Minsker, L., Joannet, E., & Renken, A. (2004).

If packed bed micro-reactor is employed for the reaction, reaction engineering model will take a similar form as that of trickle bed reactor. In packed bed micro-reactor, at low gas−liquid flow rates, possibility of existence of non-wetted portion of the bed needs to be examined and appropriately accounted for. Plug flow-type models can be employed along the length of the reactor and diffusion effects can be neglected due to smaller particle sizes. Suitable correlations to estimate liquid holdup and wetting parameters might not be available in the literature and may have to be determined from the experiments.

Applications

With their unique characteristics, micro-trickle bed reactors may find significant applications in the coming years. Such micro-reactors have been tested successfully for hydrogenation of cyclohexene (Lee, Yeong, Gavriilidis, Zapf, & Hessel, 2005) and benzene, benzalacetone, and nitrobenzene (Yeong, Gavriilidis, Zapf, & Hessel, 2003, 2004). Most of these hydrogenation reactions are exothermic in nature and involve multiple reactions. Better control on mixing and transport rates in micro-trickle bed reactors lead to some unique advantages. Many of the oxidation reactions are carried out near the explosive reaction conditions. Better transport rates of micro-trickle bed reactors offer lead to many opportunities to carry out such reactions without jeopardizing safety. Other reactions such as fluorination (de Mas, Günther, Schmidt, & Jensen, 2003) and catalytic dehydration of isopropanol (Rouge, Spoetzl, Gebauer, Schenk, & Renken, 2001) were successfully carried out in micro-reactor. Many of these reactions are still under investigation and are not yet implemented in commercial-scale operations. Emphasis on product quality and safety may encourage wider applications of micro-trickle bed reactors, especially in specialty and pharmaceutical applications.

CLOSURE

The analysis and design of trickle bed reactor is a complex task as illustrated in the various chapters of this book and it is true that a predictive model to precisely determine the size of a reactor is not possible at this stage. However, use of empirical models like pseudo-homogeneous models is also not recommended as many important issues related to mass and heat transfer coupled with catalytic reactions and hydrodynamic complexities are not adequately captured by these, especially when they have to be used for scale-up. In this book an attempt has been made to bring out all the relevant fundamental issues in the analysis of trickle bed reactors that would allow evaluation of the simultaneous effects of many micro-scale and meso-scale processes with reactions. Detailed description of general technological advantages, evaluation of hydrodynamic regimes, and estimation of design parameters are given including empirical correlations. A detailed description of reaction engineering models for simple as well as complex multistep reactions has been given along with experimental validation of the models using design parameter correlations. Further, advanced models to understand complex fluid dynamics and mass/heat transfer processes and multiscale modeling approaches which combine CFD and reaction engineering models have been discussed. This approach reduces the uncertainty associated with the use of conventional models and allows better understanding of reactor engineering and scale-up of trickle bed reactors. A few specific case studies and recent developments in monolithic and multichannel micro-trickle bed reactors are also discussed with their advantages and limitations. At this juncture, it may be useful to include some comments based on our experience of applying these models for addressing industrial needs.

We have made an attempt to provide a systematic approach to understand the various elements of hydrodynamics, interphase and intraparticle mass and heat transfer, fluid phase mixing, and reaction kinetics relevant to trickle bed reactors for single as well as multiple reactions. The discussion on key reactor engineering issues and various modeling approaches will help in selecting appropriate models and their combinations. It is important to emphasize that it is essential to correctly:

- identify and define the reactor engineering objectives,
- analyze various key issues relevant to achieving the defined objectives, and
- formulate an appropriate modeling approach/ tools, which are consistent with the set objectives.

The so-called Occam's razor provides useful tips for selecting appropriate methods/tools. Occam's razor can be stated as, "it is futile to do with more, what can be done with less." There are many instances where simple, conventional models may provide elegant and adequate solution. Even if complete solutions are not possible with simple, conventional models, conventional analysis and modeling is essential for understanding the problem correctly and for appropriate formulation of the more rigorous CFD models. Conventional flow modeling and accumulated empirical knowledge about the trickle beds must be used to get whatever useful information that can be obtained as a first step, before undertaking rigorous CFD modeling. It is, however, equally important to emphasize here the maxim that says "one should always try to make things as simple as possible (following the Occam's razor) but not simpler." It may be necessary to match the complexity of the problem with complexity of the analyzing tool. One may try to find simple solutions to complex problems, which may not be right all the time! Distinguishing the "simple" (keeping the essential aspects intact and ignoring non-essential aspects) and "simpler" (ignoring some of the crucial issues along with the non-essential issues) formulations is a very important step toward finding useful solutions to practical problems. One should have an expertise and skill to select an appropriate level of complexity of the analyzing tools to suit the set objectives. A diagnostic analysis of significance of various factors that may be contributing to specific process performance is another approach to simplify the models and select appropriate advanced models for design purpose. This is one of the most important pre-requisite for successful execution of reactor engineering projects. We hope that discussion in this book will help a reactor engineer in making appropriate selection of modeling approach and models.

Some comments on research needs and future trends in reactor engineering may be appropriate at this juncture. With the emergence of cheap, high-speed computing platforms and availability of the commercial modeling tools, computational modeling needs to be harnessed to devise best possible reactor hardware. In order to fully realize the potential of such computational modeling for better reactor engineering, it is essential to focus further research on (1) development and validation of various key physico-chemical processes occurring in trickle beds viz partial wetting, liquid distribution, capillary forces, hysteresis, inherent dynamics, non-isothermal effects, and so on and (2) a multiscale framework with which models addressing processes occurring on different spatio-temporal scales can communicate and complement each other. This list is merely suggestive since the complexity of reactive flows may greatly expand the list of issues on which further research is required. Accepting the limitations of knowledge of underlying physics and compensating these by model calibration whenever necessary, is essential to expand the applications of computational modeling to practical reactor engineering. Such experience will provide invaluable information and guidance for future development. Besides this, such applications will significantly enhance the current reactor engineering practice.

We hope that the contents of this book will be useful in developing models for trickle beds and stimulate applications of computational flow modeling to reactor engineering of trickle beds.

REFERENCES

Albers, R. E., Cybulski, A., Kreutzer, M.T., Kapteijn, F., & Moulijn, J.A. (2006). Modeling and Design of Monolith Reactors for Three-Phase Processes. In Cybulski & Moulijn (Eds.). Structured catalysts and Reactors by Cybulski and Moulin, CRC Press.

Albers, R. E., Nystrom, M., Siverstrom, M., Sellin, A., Dellve, A.-C., Andersson, U., et al. (2001). Development of a monolith-based process for H_2O_2 production: from idea to large scale implementation. *Catalysis Today, 69*, 247.

Al-Dahhan, M. H., & Dudukovic, M. P. (1995). Catalyst wetting efficiency in trickle-bed reactors at high pressure. *Chemical Engineering Science, 50*(15), 2377.

Amador, C., Gavriilidis, A., & Angeli, P. (2004). Flow distribution in different microreactor scale-out geometries and the effect of manufacturing tolerances and channel blockage. *Chemical Engineering Journal, 101*, 379–390.

Bercic, G., & Pintar, A. (1997). The role of gas bubbles and liquid slug lengths on mass transport in the Taylor flow through capillaries. *Chemical Engineering Science, 52*(21–22), 3709–3719.

Bey, O., & Eigenberger, G. (1997). Fluid flow through catalyst filled tubes. *Chemical Engineering Science, 52*, 1365.

Bhaskar, M., Valavarasu, G., Sairam, B., Balaraman, K. S., & Balu, K. (2004). Three-phase reactor model to simulate the performance of pilot-plant and industrial trickle-bed reactors sustaining hydrotreating reactions. *Industrial and Engineering Chemistry Research, 43*, 6654–6669.

Bretherton, F. P. (1961). The motion of long bubbles in tubes, *Journal of Fluid Mechanics, 10*, 166–188.

Charpentier, J. C., & Favier, M. (1975). Some liquid holdup experimental data in trickle-bed reactors for foaming and non-foaming hydrocarbons. *AIChE Journal, 21*, 1213e1218.

Chowdhury, R., Pedernera, E., & Reimert, R. (2002). Trickle-bed reactor model for de-sulfurization and dearomatization of diesel. *AIChE Journal, 48*, 126.

Crezee, E., Barendregt, A., Kapteijn, F., & Moulijn, J. A. (2001). Carbon coated monolithic catalysts in the selective oxidation of cyclohexanone. *Catalysis Today, 69*, 283–290.

Cybulski, A., & Moulijn, J. A. (Eds.). (2006). *Structured catalysts and reactors*. Boca Raton, Florida: CRC Taylor Francis.

Debellefontaine, H., & Foussard, J. N. (2000). Wet air oxidation for the treatment of industrial wastes. Chemical aspects, reactor design and industrial applications in Europe. *Waste Management, 20*, 15–25.

de Mas, N., Günther, A., Schmidt, M. A., & Jensen, K. F. (2003). Microfabricated multiphase reactors for the selective direct fluorination of aromatics. *Industrial Engineering and Chemical Research, 42*(4), 698–710.

Eftaxias, A., Font, J., Fortuny, A., Giralt, J., Fabregat, A., & Stüber, F. (2001). Kinetic modelling of catalytic wet air oxidation of phenol by simulated annealing. *Applied Catalysis B: Environmental, 33*, 175–190.

Eftaxias, A., Larachi, F., & Stüber, F. (2003). Modelling of trickle bed reactor for the catalytic wet air oxidation of phenol. *The Canadian Journal of Chemical Engineering, 81*, 784–794.

Ehrfeld, W., Hessel, V., & Löwe, H. (2000). *Microreactors: New Technology for Modern Chemistry*. Weinheim: Wiley-VCH.

Fortuny, A., Bengoa, C., Font, J., Castells, F., & Fabregat, A. (1999). Water Pollution Abatement by Catalytic Wet Air Oxidation in a Trickle Bed Reactor. *Catalysis Today, 53*, 107–114.

Fukushima, S., & Kusaka, K. (1977). Liquid Phase volumetric and mass transfer coefficient, and boundary of hydrodynamic flow region in packed column with cocurrent downward flow. *Journal of Chemical Engineering of Japan, 10*(6), 468.

Goto, S., Levec, J., & Smith, J. M. (1977). Trickle-bed oxidation reactors. *Catalysis Reviews, 15*(1), 187–247.

Goto, S., & Smith, J. M. (1975). Trickle-bed reactor performance. Part-I. Holdup and mass transfer effects. *AIChE Journal, 21*, 706–713.

Goto, S., & Smith, J. M. (1978). Performance of slurry and trickle-bed reactors: application to sulfur dioxide removal. *AIChE Journal, 24*, 286.

Gunjal, Prashant R., Kashid, Madhavanand N., Ranade, Vivek V., & Chaudhari, Raghunath V. (2005a). Hydrodynamics of trickle-bed reactors: experiments and CFD modeling. *Industrial and Engineering Chemistry Research, 44*, 6278–6294.

Gunjal, Prashant R., Ranade, Vivek V., & Chaudhari, Raghunath V. (2005b). Computational study of a single-phase flow in packed beds of spheres. *AIChE Journal, 51*(2), 365.

Hatziantoniou, V., & Andersson, B. (1984). The segmented two phase flow monolithic catalyst reactor: an alternative for liquid phase hydrogenations. *Industrial and Engineering Chemistry Fundamentals, 23*, 82–88.

Heiszwolf, J., Engelvaart, L., van der Eijnden, M., Kreutzer, M., Kapteijn, F., & Moulijn, J. (2001). Hydrodynamic aspects of the monolith loop reactor. *Chemical Engineering Science, 56*, 805–812.

Hessel, V., Löb, P., & Löwe, H. (2005). Development of microstructured reactors to enable organic synthesis rather than subduing chemistry. *Current Organic Chemistry, 9*, 765.

Irandoust, S., & Anderson, B. (1988). Monolithic catalysts for nonautomobile applications. *Catalysis Reviews — Science and Engineering, 30*(3), 341–392.

Irandoust, S., & Andersson, B. (1989). Liquid film in Taylor flow through a capillary. *Industrial and Engineering Chemistry Research, 28*, 1684–1688.

Jensen, K. F. (2001). Microreaction engineering — is small better? *Chemical Engineering Science, 56*, 293–303.

Jiang, Y., Khadilkar, Mohan R., Al-Dahhan, Muthanna H., & Dudukovic, Milorad P. (2001). CFD modeling of multiphase flow distribution in catalytic packed bed reactors: scale down issues. *Catalysis Today, 66*, 209–218.

Khadilkar, M. R., Wu, Y. X., Al-Dahhan, M. H., & Dudukovic, M. P. (1996). Comparison of trickle bed and upflow reactor performance at high pressure: Model prediction and experimental observations. *Chemical Engineering Science, 51*(10), 2139.

Kiwi-Minsker, L., Joannet, E., & Renken, A. (2004). Loop reactor staged with structured fibrous catalytic layers for liquid-phase hydrogenations. *Chemical Engineering Science, 59*, 4919–4925.

Kiwi-Minsker, L., & Renken, A. (2005). Microstructured reactors for catalytic reactions. *Catalysis Today, 110*(1–2), 2–14.

Kreutzer, M. T. (2003). *Hydrodynamics of Taylor Flow in Capillaries and Monolith Reactors.* Delft, The Netherlands: Delft University Press.

Kreutzer, M. T., Du, P., Heiszwolf, J. J., Kapteijn, F., & Moulijn, J. A. (2001). Mass transfer characteristics of three-phase monolith reactors. *Chemical Engineering Science, 56*, 6015–6023.

Kreutzer, M. T., Kapteijn, F., Moulijn, J. A., Kleijn, C. R., & Heiszwolf, J. J. (2005). Inertial and interfacial effects on pressure drop of Taylor flow in capillaries. *AIChE Journal, 51*(9), 2428–2440.

Laborie, S., Cabassud, C., Durand-Bourlier, L., & Laine, J. M. (1999). Characterization of gas–liquid two-phase flow inside capillaries. *Chemical Engineering Science, 54*, 5723.

Lee, J. W., Yeong, K. K., Gavriilidis, A., Zapf, R., & Hessel, V. (2005). In *Proceedings of the international conference on microreaction technology* (IMRET8). Atlanta, USA.

Levec, J., & Smith, J. M. (1976). Oxidation of acetic acid solutions in a trickle-bed reactor. *AIChE Journal, 22*, 159–168.

Liu, H., Zhao, J. D., Li, C. Y., & Ji, S. F. (2005). Conceptual design and CFD simulation of a novel metal-based monolith reactor with enhanced mass transfer. *Catalysis Today, 105*, 401–406.

Losey, M. W., Jackman, R. J., Firebaugh, S. L., Schmidt, M. A., & Jensen, K. F. (2002). Design and fabrication of microfluidic devices for multiphase mixing and reaction. *Journal of Microelectromechanical Systems, 11*, 709–717.

Losey, M. W., Schmidt, M. A., & Jensen, K. F. (2001). Microfabricated multiphase packed-bed reactors: characterization of mass transfer and reactions. *Industrial and Engineering Chemistry Research, 40*, 2555–2562.

Mathew, S. P. (2001). *Catalytic Hydrogenation of Aromatic Ketones in Multiphase Reactors: Catalysis and Reaction Engineering Studies. Ph.D. Thesis.* India: University of Pune.

Mazzarino, I., & Baldi, G. (1987). Liquid phase hydrogenation on a monolith catalyst. In B. Kulkarni, R. Mashelkar, & M. Sharma (Eds.), *Recent trends in chemical reaction engineering* (pp. 181). New Delhi: Wiley Eastern.

Mills, P. L., & Dudukovic, M. P. (1979). A dual-series solution for the effectiveness factor of partially wetted catalysts in trickle-bed reactors. *Industrial and Engineering Chemistry Fundamentals, 18*, 139–149.

Mills, P. L., & Dudukovic, M. P. (1984). A comparison of current models for isothermal trickle-bed reactors. *ACS Symposium Series, 237*, 37−59.

Mills, P. L., Ramachandran, P. A., & Chaudhari, R. V. (1992). Multiphase reaction engineering for fine chemicals and pharmaceuticals. *Reviews in Chemical Engineering, 8*, 1−176.

Mueller, G. E. (1991). Prediction of radial porosity distribution in randomly packed fixed beds of uniformly sized spheres in cylindrical containers. *Chemical Engineering Science, 46*, 706.

Pedernera, E., Reimert, R., Nguyen, N. L., & van Buren, V. (2003). Deep desulfurization of middle distillates: process adaptation to oil fractions' compositions. *Catalysis Today, 79−80*, 371−381.

Pintar, A., & Levec, J. (1994). Catalytic liquid-phase oxidation of phenol aqueous solutions. A kinetic investigation. *Industrial and Engineering Chemistry Research, 33*, 3070−3077.

Rajashekharam, M. V., Jaganathan, R., & Chaudhari, R. V. (1998). A trickle-bed reactor model for hydrogenation of 2,4-dinitrotoluene: experimental verification. *Chemical Engineering Science, 53*(4), 787−805.

Ramachandran, P. A., & Chaudhari, R. V. (1983). *Three Phase Catalytic Reactors.* New York: Gordon and Breach.

Rodrýguez, M. A., & Ancheyta, J. (2004). Modeling of hydrodesulfurization (HDS), hydro-denitrogenation (HDN), and the hydrogenation of aromatics (HDA) in a vacuum gas oil hydrotreater. *Energy and Fuels, 18*, 789−794.

Rouge, A., Spoetzl, B., Gebauer, K., Schenk, R., & Renken, A. (2001). Microchannel reactors for fast periodic operation: the catalytic dehydration of isopropanol. *Chemical Engineering Science, 56*(4), 1419−1427.

Roy, S., Bauer, T., Al-Dahhan, M., Lehner, P., & Turek, T. (2004a). Monoliths as multiphase reactors: a review. *AIChE Journal, 50*, 2918−2938.

Roy, S., Heibel, A. K., Liu, Wei, & Boger, T. (2004b). Design of monolithic catalysts for multiphase reactions. *Chemical Engineering Science, 59*(5), 957−966.

Satterfield, C. N. (1970). *Mass Transfer in Heterogeneous Catalysis.* Cambridge, MA: MIT Press.

Satterfield, C. N., & Ozel, F. (1977). Some characteristics of two-phase flow in monolithic catalyst structures. *Industrial and Engineering Chemistry Fundamentals, 16*, 61−67.

van Baten, J. M., & Krishna, R. (2004). CFD simulations of mass transfer from Taylor bubbles rising in circular capillaries. *Chemical Engineering Science, 59*, 2535−2545.

van Baten, J. M., & Krishna, R. (2005). CFD simulations of wall mass transfer for Taylor flow in circular capillaries. *Chemical Engineering Science, 60*, 1117−1126.

Vandu, C. O., & Krishna, R. (2004). Volumetric mass transfer coefficients in slurry bubble columns operating in the churn-turbulent flow regime. *Chemical Engineering and Processing, 43*, 987−995.

Vandu, C. O., Liu, H., & Krishna, R. (2005). Mass transfer from Taylor bubbles rising in single capillaries. *Chemical Engineering Science, 60*, 6430−6437.

Wada, Y., Schmidt, M. A., & Jensen, K. F. (2006). Flow distribution and ozonolysis in gas−liquid multichannel microreactors. *Industrial and Engineering Chemistry Research, 45*, 8036−8042.

Warnier, M. J. F., De Croon, M. H. J. M., Rebrov, E. V., & Schouten, J. C. (2010). Pressure drop of gas−liquid Taylor flow in round micro-capillaries for low to intermediate Reynolds numbers. *Microfluidics and Nanofluidics., 8*(1), 33−45.

Wilhite, B. A., Moreno, A., & Kim, D. (2008). Process Intensification in Microreactors. *Patent pending, #12/263,637.*

Yeong, K. K., Gavriilidis, A., Zapf, R., & Hessel, V. (2003). Catalyst preparation and deactivation issues for nitrobenzene hydrogenation in a falling film reactor. *Catalysis Today, 81*, 641−651.

Yeong, K. K., Gavriilidis, A., Zapf, R., & Hessel, V. (2004). Experimental studies of nitrobenzene hydrogenation in a microstructured falling film reactor. *Chemical Engineering Science, 59*, 3491−3494.

A	area or concentration of gas A
A^*	concentration of gas A in liquid in equilibrium with gas
a	interfacial area per unit volume
a_{GL}, a_B	specific gas–liquid interfacial area
a_{LS}	specific liquid–solid interfacial area
a_p	external area of the particles per unit volume of reactor
a_w	wetted external area of particles per unit volume of reactor
a_s, a_v	surface area of solid per unit volume of bed
a_W	reactor wall interfacial area per unit volume
B_{li}	initial concentration of B in liquid phase
B_s	concentration of B on the catalyst surface
C	concentration
$C_{1\varepsilon}, C_{2\varepsilon}, C_\mu$	parameters of the $k–\varepsilon$ model
C_D	drag coefficient
C_p^*, C_p	heat capacity
D	bed diameter or dispersion coefficient or molecular diffusivity
D_e	effective diffusivity
D_{eff}	effective diffusion coefficient
D_k	Knudsen diffusion coefficient
D_{eA}	effective diffusivity in the pores of the catalyst
D_{eg}	axial dispersion coefficient of gas phase
D_{el}	axial dispersion coefficient of liquid phase
D_{ax}	axial dispersion coefficient
D_{H_2}	diffusivity of hydrogen
d_h	hydraulic diameter, Krischer–Kast hydraulic diameter, $$d_h = d_P \sqrt[3]{\frac{16\varepsilon^3}{9\pi(1-\varepsilon)^2}}$$
d_p	particle diameter, equivalent pellet diameter
d_{ps}	equivalent spherical diameter, i.e., the diameter of the sphere having the same volume as that of particle (used in Table 1 of Chapter 2)
d_1, d_2	maximum and minimum diameter of particle with liquid film
d_c	capillary diameter

Trickle Bed Reactors. DOI: 10.1016/B978-0-444-52738-7.10001-4

E_{diff}	Activation energy of diffusion
E_i	Activation energy of reaction i
E_o	activation energy
$E(t)$	exit age distribution
E_1, E_2	Ergun's constant
F	external or interphase force or marker function
F_D	drag force
$F_{k,r}$	interface momentum exchange source for phases k and r
F_{SF}	continuum surface force
F_{GL}, F_{GS}, F_{LS}	interphase momentum exchange source between gas–liquid, gas–solid, and liquid–solid phases, respectively
f	degree of wetting
f_t	total frictional factor
f_w	fraction of external area wetted by liquid, a_w/a_p
f_s	fraction of catalyst wetted by static zone
G	gas mass velocity or generation of turbulent kinetic energy
G_G, G_L	superficial mass velocity of gas and liquid, respectively
g	acceleration due to gravity
H	height of packing/bed or enthalpy
H_2	hydrogen
He	Henry's constant
H_A	solubility coefficient of A, defined as A_g/A_l at equilibrium
h	heat transfer coefficient or enthalpy
h_{GS}, h_{LS}	gas–particle and liquid–particle heat transfer coefficients, respectively
h_{GW}, h_{LW}	gas–wall and liquid–wall heat transfer coefficients, respectively
J_o	zero-th order Bessel function
j_G, j_L	gas and liquid superficial velocities, respectively
K_{ad}	adsorption coefficient
k	kinetic energy or relative permeability
k_s	liquid-to-catalyst mass transfer coefficient
k_L	liquid film mass transfer coefficient
k_{GL}, k_{LS}, k_{GS}	gas–liquid, liquid–solid, and gas–solid mass transfer coefficients, respectively
k_{mono}	backward pseudo-first order rate constant of dearomatization reaction
k_t	turbulent thermal conductivity
k_f	thermal conductivity of fluid
l_p	length of the particle
L	liquid phase mass velocity or length

m	order of reaction
M_f	maldistribution factor
M_G, M_L	molecular weight of gas and liquid phases, respectively
N_1	first normal shear stress
n	surface normal, power law index
nr	number of reactions
P	pressure
Pr_t	turbulent Prandtl number
P_v	vapor pressure of liquid
P_c	capillary pressure
P_i	Thiele modulus of reaction i
q	heat flux, heat transfer rate, flow rate
q_B	stoichiometric ratio
q_{c1}	Dimensionless parameter in Equation (13) of chapter 3, $$q_{c1} = \alpha_{c1} \frac{D_{eff,A}}{D_{eff,C}} \frac{A^*}{B_1}$$
q_{c2}	Dimensionless parameter in Equation (13) of chapter 3, $$q_{c2} = \alpha_{c2} \frac{D_{eff,A}}{D_{eff,C}} \frac{A^*}{B_1}$$
Q	flow rate
Q_{Li}	flow rate of liquid in i-th channel
Q_{mean}	mean flow rate
r	radial coordinate, reaction rate
r^*	non-dimensional distance from container wall
R	reaction rate or radius or universal gas constant
S	source
S_{pk}	rate of mass transfer from p-th phase to k-th phase
S_p	particle geometric area
Sh'_A	Sherwood number for species A (Figure 5 of Chapter 3), $Sh'_A = k_{eff} R / D_{eff,a}$ where $k_{eff} = \left[\dfrac{1}{k_{La}(a_L/a_S)} + \dfrac{1}{k_{sa}} \right]^{-1}$
T	temperature
t	time
u, v, w	velocity, in x, y, and z directions, respectively
U, U_s, U_0, u	superficial velocity
u_i	interstitial velocity
u'	fluctuating velocity
V	volume of reactor
V_G, V_L	gas and liquid mass velocities, respectively
w	catalyst mass per unit volume of reactor
x	mass fraction

X	mole fraction
X_G, X_L	modified Lockhart–Martinelli number
x	coordinate direction
y	coordinate direction
z	coordinate direction

Greek notations

α	phase volume fraction
β	saturation or permeability of porous media
$\beta_{t,tr}$	total liquid saturation (static + dynamic volume of liquid holdup/void volume)
β_d	dynamic liquid saturation
β_e	external liquid saturation
β_L, β_l	liquid saturation
β_{nc}	non-capillary liquid saturation
δ_{ij}	Kronecker delta function
δ	pressure drop per unit length for single phase flow
ΔP	pressure drop
ε	porosity, voidage, turbulent energy dissipation rate
ε_k	volume fraction/holdup of k-th phase
ε_G	gas holdup
ε_L	liquid holdup
ε_{Ld}	dynamic liquid holdup
ε_S	solid holdup
ε_L^o, ε_{Ls}	static liquid holdup
ε_B	bed voidage
ε_P	catalyst pellet porosity
ϕ, ϕ_1	Thiele modulus or general variable
ϕ_0	value of Thiele modulus based on inlet conditions (or bulk conditions for a semi-batch reactor)
ϕ_p	particle sphericity
φ	shape factor
Ω	local rate of chemical reaction
η, η_c	effectiveness factor or wetting efficiency
η_o	overall effectiveness factor
η_d	effectiveness factor for a dry catalyst pellet
η_l, η_w	effectiveness factor in wetted region
η_{CE}	external contacting efficiency
κ	curvature
λ_ι	solubility
λ_e^o	bed thermal conductivity
λ	thermal conductivity
$\dot{\gamma}$	shear rate

$\dot{\gamma}_W$	shear rate at wall
γ_1	Arrhenius number for the first reaction $\gamma_1 = E_1/RT_1$
γ_2	Arrhenius number for the first reaction $\gamma_2 = E_2/RT_1$
γ_3	Dimensionless heat of adsorption, $\gamma_3 = (-\Delta H_{ads})/RT_1$
γ_4	Dimensionless activation energy for diffusion, $\gamma_4 = E_{diff}/RT_1$
γ_5	Dimensionless heat of solution, $\gamma_5 = (-\Delta H_{sol})/RT_1$
λ_{eff}	effective thermal conductivity
Λ_r	radial thermal conductivity
μ	molecular viscosity
μ_{eff}	effective viscosity
ν	kinematic viscosity of the fluid
$\nu_b, \nu_{c1}, \nu_{c2}, \nu_e$	Stoichiometric coefficients
ν_n	molar gas volume
ν_t	turbulent kinematic viscosity
θ_t	contact angle
θ	dimensionless exit temperature (T/T_i)
θ	Dimensionless temperature, $\theta = T/T_1$
ρ	density of fluid
ρ_B	catalyst bulk density
σ	surface tension
$\sigma_\kappa, \sigma_\varepsilon$	parameters of the $k-\varepsilon$ model
τ	viscous stress tensor, residence time, tortuosity
τ	shear stress (Table 1, Chapter 2)
τ_s	solid stress tensor
τ_w	wall shear stress
ω	vorticity
ΔH	heat of reaction
ΔH_{ads}	Heat of adsorption
ΔH_{sol}	Heat of solution
ψ_1	Dimensionless rate, $\psi_1 = P_1 a^m b^n \exp\left[-\gamma_1\left(\frac{1}{\theta}-1\right)\right]$ power law kinetics
ψ_1	Dimensionless rate, $\psi_1 = P_1 \dfrac{ab\exp\left[(-\gamma_1+\gamma_4)(\frac{1}{\theta}-1)\right]}{Den^u}$ L-H kinetics
ψ_2	Dimensionless rate, $\psi_2 = P_2 a^p c^q \exp\left[-\gamma_2\left(\frac{1}{\theta}-1\right)\right]$ power law kinetics

ψ_2 Dimensionless rate, $\psi_2 = P_2 \dfrac{ac \exp\left[(-\gamma_2 + \gamma_4)\left(\frac{1}{\theta} - 1\right)\right]}{Den^u}$

L-H kinetics

Subscripts

α	phase alpha
A	component A
b	bulk
d	dynamic zone
eff	effective
F	frictional
G, g	gas
g	dry zone, gas phase
in	inlet
i	i-th direction
j	j-th direction
k	k-th direction, k-th phase
L, l	liquid
out	outlet
Obs	observed
p, P	particle
S, s	solid phase, static zone, Slug
t	tangential or turbulent or at transition
ν_e	viscoelastic liquid phase
ν_i	viscoinelastic liquid phase
w	wall

Abbreviations

ACPH	Acetophenone
Ar_S	aromatics sulfur compound
BCC	body centered cubic
CDA	cyclododecane
CDD	cyclododecadiene
CDE	cyclododecene
CDT	1,5,9-cyclododecatriene
CFD	computational fluid dynamics
CHET	1-cyclohexylethanol
CHMK	cyclohexylmethylketone
CT	computed tomography
DCPD	dicyclopentadiene
Di_{Ar}	diaromatics

ETBE	ethylbenzene
ETCH	ethylcyclohexane
FCC	face centered cubic
H_2	hydrogen
LHSV	liquid hourly space velocity
MRI	magnetic resonance imaging
$mono_{Ar}$	monoaromatics
PHET	1-phenylethanol
r	reactant
SC	simple cubic
Tri_{Ar}	triaromatics

Dimensionless numbers

Bodenstein number	$Bo = \dfrac{Uz}{D_{ax}}$
Ca	capillary number, $U\mu/\sigma$
Dimensionless pressure drop	$\psi_\alpha = \dfrac{\Delta P}{\Delta z}/(\rho_\alpha g + 1)$
Froude number	$Fr = \dfrac{U^2}{d_p g}$
Ga^*	$Ga^* = d_p^3 \rho_L (\rho_L g + \Delta P/\Delta Z)/\mu_L^2$, modified Galileo number, dimensionless
Galileo number	$Ga_\alpha = \dfrac{d_p^3 g \rho_\alpha^2}{\mu_\alpha^2}, \quad Ga_\alpha = \dfrac{d_p^3 g \rho_\alpha^2}{\mu_\alpha^2 (1 - \varepsilon)^3}$
Graetz number	$Gz = \dfrac{\Psi}{Re\,Pr}$
Kapitza number	$Ka = \dfrac{\sigma_L^3 \rho_L}{\mu_L^4 g}$
Liquid Peclet number ($Re_l Pe_l$)	$Pe_l = \dfrac{L C_{pl} d_p}{k_l}$
Modified Eötvos number	$E\ddot{o} = \dfrac{\rho_L g d_p^2 \phi^2 \varepsilon^2}{\sigma_L (1 - \varepsilon)^2}$
Modified Lockhart–Martinelli	$X_L = \dfrac{1}{X_G} = \dfrac{U_L}{U_G}\left(\sqrt{\dfrac{\rho_L}{\rho_G}}\right)$
Nusselt number	$Nu = \dfrac{hd}{k}$

Particle Reynolds number	$Re_P = \dfrac{\rho U_0 d_P}{\mu}$
Peclet number	$Pe_\alpha = \dfrac{U_\alpha L}{D_{ax}}$
Prandtl number	$Pr = \dfrac{\mu C_p}{k}$
Reynolds number	$Re = \dfrac{\rho U d}{\mu}$
Schmidt number	$Sc = \dfrac{\mu}{\rho D}$
Sherwood number	$Sh_\alpha = \dfrac{k_\alpha d_h}{D_\alpha}$
Weber number	$We_\alpha = U_\alpha^2 d_P \rho_\alpha / \sigma_\alpha$

Author Index

A

Achenbach, E., 64
Acrivos, A., 127, 129
Afacan, A., 146, 188
Aitamma, J., 101
Akehata, T., 64
Akgerman, A., 90
Al-Dahhan, M. H., 4, 30, 39, 44, 52, 54, 57, 59, 93, 98, 101, 121−123, 146, 175, 182, 188, 193, 219, 225, 235
Alexander, P., 128
Alopaeus, V., 101
Ananth, M. S., 46, 160
Arntz, D., 98
Attou, A., 32, 36−39, 146−148
Avraam, D. G., 95

B

Baker, O., 34
Bakos, M., 26
Baldi, G., 29, 32, 46, 55, 57, 66, 160−161, 246
Ballard, J. H., 183
Bansal, A., 33−35
Bansal, M., 63
Bartelmus, G., 33, 46
Bartholdy, J., 187
Baussaron, L. C., 52−54
Beaudry, E. G., 87, 98
Benaissa, M., 94
Benkrid, K., 43
Benyahia, F., 142
Bercic, G., 94, 242
Bergault, I., 92, 95, 99
Bernard, J. R., 64, 187
Bernard, R. S., 127
Bertucco, A., 95
Bhatia, S. K., 177−178
Bischoff, K. B., 82, 106
Bocharov, A. A., 190
Boelhouwer, J. G., 43, 47, 64−65, 86, 109, 162, 193

Borda, M., 185, 187−188
Boyer, C., 52
Bramley, A. S., 128
Briens, C. L., 64
Brzić, D., 108
Buffham, B. A., 63
Burghardt, A., 33, 46−47, 52−54

C

Calis, H. P. A., 120, 128−129
Canu, P., 95
Carbonell, R. G., 32, 36, 42, 51, 87, 146−147
Carman, P. C., 144
Cassanello, M., 55, 61, 63, 191
Castellari, A. T., 94, 99, 101, 193
Cechini, J., 193
Cechini, J. O., 99
Charpentier, J. C., 26, 29, 32, 37−38, 42, 44, 49, 55, 58, 60, 250
Chaudhari, R. V., 5, 23, 27, 78, 80−83, 88, 92, 95, 104−106, 108−111, 123, 132, 150−151, 164, 174−175, 188, 192, 204, 214, 216, 223, 225
Chen, J., 75
Chhabra, R. P., 27, 129
Choudhury, D., 131
Chou, T. S., 32−33, 37, 57, 58, 185
Chuang, K. T., 188
Chu, C. F., 64
Chung, K. T., 146
Clements, L. D., 49
Colakyan, M., 175
Collins, G. M., 90
Colombo, A. J., 51
Comiti, J., 27, 130
Crine, M., 146, 185
Cukierman, A. L., 61, 191

D

Danckwerts, P. V., 60
Dautzenberg, F. M., 128

Davis, H. T., 127
Delaunay, C. B., 58
Delikat, E. P., 34, 185–186
Delmas, H., 52, 88, 92, 94–95, 174–175
Del, Pozo, M., 64
Devetta, L., 95
Dhole, S. D., 129, 138
Dietz, A., 95
Dillon, P. O., 33
Dixon, A. G., 128–129, 142–143
Donohue, T. J., 121, 123, 125
Drinkenburg, A. A. H., 43, 58, 64, 86, 109, 162, 193
Dudukovic, M. P., 26, 30, 32, 35, 39, 44, 52–54, 57, 78, 87, 92–93, 98, 101, 121–123, 146–147, 175, 182, 188, 191, 193, 214–215, 219, 225
Durst, F., 127

E

Effron, E., 61, 63, 191
Eftaxias, A., 232–234
El-Hisnawi, A. A., 53–54, 97–98
El-Hisnawi, A. E., 52
Ellenberger, J., 191
Ellman, M. J., 42, 49–50
Emig, G., 82, 95
Emig, Merchan. G., 82, 84–85
Ergun, S., 35–36, 43, 138, 140, 144, 146, 159–160
Eswaran, V., 129
Euzen, J. P., 94

F

Favier, M., 32, 37–38, 44, 55
Ferschneider, G., 32, 36–39, 146–148
Fogler, H. S., 146
Foumeny, E. A., 142
Fox, R., 185
Freund, H., 127–128
Fujiyoshi, K., 65
Fukai, J., 150
Funk, G. A., 185, 188–189, 199

G

Gabarain, L., 193
Gabitto, J. F., 185
Galloway, T. R., 59–60

Gamborg, M. M., 183
Gaskey, S. W., 94
Gence, J. M., 27
Gerhard, H., 33
Germain, A. H., 63, 87
Gerth, L., 66
Gianetto, A., 29, 32, 55, 57, 78
Gilbert, J. B., 102
Gladden, L. F., 51, 53, 121, 128, 132, 186
Goff, P., 26
Goto, S., 59, 87, 92, 108, 174, 231
Greenfield, P. F., 42
Grosser, K., 32–33, 35, 37–39, 146–148
Gulijk, C. V., 191
Gunjal, P. R., 28, 33, 37, 40–41, 45–46, 62, 122–124, 132–141, 146, 150–151, 154–159, 161–162, 164, 188–189, 192, 202, 225–226, 229–231
Guo, J., 95–97, 101
Gutsche, R., 108

H

Haas, R., 98
Haas, T., 127
Halken, C., 187
Hanika, J., 63, 89–91, 95, 108, 176–177, 193
Hanratty, P. J., 191
Hanratty, T. J., 27
Hansen, J. A., 187
Harold, M. P., 87, 97, 177, 185, 189
Hashimoto, K., 64–66
Haure, P. M., 99–101, 108, 192–193
Hawley, M. C., 144
Helwick, J. A., 33
Henry, H. C., 102
Herrmann, U., 95
Herskowitz, M., 26, 51–52, 87, 98, 174–175, 185, 187
Hessari, F. A., 177–178
High, Fill, W., 59
Higler, A. P., 191
Hill, R. J., 127, 130
Hines, J. E., 183
Hirai, S., 128
Hirose, T., 58–59
Hirt, C. W., 150
Hitaka, Y., 64
Hochman, J., 61, 63, 191
Hoffmann, U., 94, 99
Hofmann, H., 33, 39, 82, 97–98, 102

Holub, R. A., 32−34, 38−39, 44−45, 49−50, 146−147, 160−161
Hook, B. D., 90
Ho, T. C., 90, 176
Huang, T. C., 64, 92, 98
Huang, X., 65
Huber, F., 176
Hudgins, R. R., 100, 192−193
Huet, F., 27, 130
Hu, R., 176
Hu, X., 126

I

Ida, Y., 63
Iliuta, I., 55, 60−61, 94, 101
Iliuta, M. C., 94
Interthal, W., 127

J

Jaffe, S. B., 63, 97, 202−203
Jaganathan, R., 88, 174−175, 223
Jaroszynski, M., 46, 53
Jensen, B. N., 183
Jensen, K. F., 8, 249, 251
Jiang, Y., 93, 95, 97, 100, 122−123, 146, 148−149, 188, 225
Johns, M. L., 128
Jolls, K. R., 27
Jongen, T., 131
Julcour, C., 52, 54, 88, 95, 174−175
Julcour-Lebigue, 52

K

Kader, B., 131
Kamperman, A. A., 191
Kang, B. C., 92, 98
Kan, K. M., 42
Kapoor, V. K., 61, 191
Karabelas, A. J., 33
Kashid, M. N., 33, 123, 192, 225
Kashiwa, B. A., 146
Kaur, N., 63
Kawase, Y., 57−59
Kenney, C. N., 87
Khadilkar, M. R., 52, 93−94, 98, 100, 122−123, 146, 175, 188, 193−194, 225, 233
Kheshgi, H. S., 176

Kim, D. H., 177
Kim, S. E., 131
Kim, Y. G., 177
Kireenkov, V. V., 190−191
Kirillov, V. A., 176, 179, 190
Klemm, E., 176
Kobayashi, S., 63
Koch, D. L., 127
Kocis, G. R., 90
Kolodziej, A., 46, 53
Koptyug, I. V., 176, 179
Koros, R. M., 89
Korsten, H., 94, 99
Kothe, D. B., 150
Krausova, J., 176
Krieg, D. A., 33
Krishna, R., 1, 29, 191, 243−244
Kroll, D. M., 127
Kufner, R., 97
Kulikov, A. V., 124
Kunugita, E., 63
Kusaka, K., 26, 30, 55, 219−220
Kushiyama, S., 63
Kutovsky, Y. E., 127
Kuzin, N. A., 176, 190

L

Ladd, A. J. C., 127
Lakota, A., 57−58, 87
Lakota, A., 57−58, 87
Lammers, P., 179
Lamine, A. S., 64, 66−67, 187
Lange, R., 55, 63, 108, 193
Larachi, F., 35, 44, 49−50, 55, 57, 60, 78, 101, 181, 232
Latifi, M. A., 27, 33, 57−58
Launder, B. E., 131
Laurent, A., 35, 38, 42, 44, 49, 55, 57−58
Lee, H. H., 88
Lee, J. K., 193
Lefebvre, A. G., 63, 87
Legall, H., 66
Lemcoff, N. O., 185
Le, Mehaute, A., 58
Levec, J., 57−58, 94, 144, 181, 231−232
Levec, L., 87
L'Homme, G. A., 63, 87, 146, 185
Liu, G., 175−176, 239−240
Logtenberg, S. A., 128−129
Lopes, R. J. G., 150−154

Loser, T., 7, 120
Lukjanov, B. N., 176
Luss, D., 33, 57, 94
Lutran, P. G., 34, 185–187
Lysova, A. A., 91–92

M

Machac, I., 27
Magnico, P., 128, 138
Maier, R. S., 127–128, 138
Mantle, M. D., 121, 128, 132, 146
Mao, Z., 35
Marcandelli, C., 64, 187–189
Marchot, P., 146, 185
Martinez, O. M., 61, 191
Martin, J. J., 129
Mathew, S. P., 88, 175, 216–222
Matsuura, A., 62, 64
McCabe, W. L., 129
McCready, M. J., 33, 64
McHyman, J., 150
McManus, R. L., 189–190
Mears, D. E., 61, 102, 191
Megaridis, C. M., 150
Mehta, D., 144
Metaxas, K. C., 175
Mewes, D., 7, 120
Michell, R. W., 63
Midoux, N., 27, 29, 35, 38, 42–44, 49, 55, 57–58
Millies, M., 7, 120
Mills, P. L., 35, 52–54, 78, 81, 87, 98, 214–215
Missen, R. W., 54
Miyatake, O., 150
Mi, Z., 175
Møller, L. B., 187–188
Monaghan, J. J., 150
Monrad, C. C., 129
Montagna, A. A., 191
Montillet, A., 27, 130
Morita, S., 87
Mori, Y., 59
Morris, C., 90
Morsi, B. I., 29, 38, 48, 55, 57
Mosseri, S., 87, 174–175
Mueller, G. E., 121–122, 124, 145, 228
Muntean, O., 61
Muroyama, K., 64–65, 67
Muthanna, J. G., 95
Myers, J. E., 26, 64–65, 185

N

Naderifar, A., 58
Nagata, S., 65
Nandakumar, K., 146, 188
Natarajan, R., 129
Nelson, P. A., 59–60
Ng, K. M., 33–34, 37–39, 64, 87, 109, 185–186, 189
Nichols, B. D., 150
Nigam, K. D. P., 31–32, 61, 191
Nijemeisland, M., 128–129, 142–143
Nijenhuis, J., 128

O

Oliveros, G., 191
Otake, T., 63
Ozel, F., 235, 240

P

Padial, N. T., 146
Paikert, B. C., 128
Pant, H. J., 61
Parmon, V. N., 124
Papayannakos, N. G., 175
Piepers, H. W., 43, 64, 86, 109, 162, 193
Pintar, A., 94, 232, 242
Pistek, R., 176
Polaert, I., 126
Poulikakos, D., 150
Prost, C., 26
Purwasasmita, M., 55

Q

Qiao, S., 126
Quinta-Ferreira, Rosa M., 150–154

R

Rajashekharam, M. V., 87–88, 92–93, 95, 100, 105, 174–175, 178, 223–224
Ramachandran, P. A., 5, 27, 32, 39, 60, 78, 80, 82–83, 87, 93, 104, 106, 147, 191, 204, 214, 216
Ranade, V. V., 33, 123–124, 130, 132, 142, 146, 150–151, 164, 188, 192, 202, 225–226, 229–231
Ranz, W. E., 64

Rao, D. P., 115, 175
Rao, M. S., 175
Rao, V. G., 46, 48, 58, 160−161
Rathor, M. N., 60, 63
Rauenzahn, R. M., 146
Ravindra, P. V., 94, 175
Reyes, S. C., 176
Rider, W. J., 150
Ring. Z. E., 54
Riopelle, J. E., 183
Rode, S., 43
Roininen, J., 96, 101
Rstka, J. H., 89
Rudman, M., 150
Ruzicka, V., 89, 176

S

Saez, A. E., 36, 42, 147, 161
Sagdeev, R. Z., 124
Sai, P. S. T., 32, 37−38, 42, 226
Sangani, A. S., 127
Saroha, A. K., 31−32, 61, 191
Sato, Y., 32, 59
Satterfield, C. N., 5, 26, 31, 52, 58−59, 78, 90, 93, 200, 214, 235, 240
Saxena, A. K., 61, 191
Scarsdale, N. Y., 183
Scheidegger, A. E., 36
Schmidt, P. C., 8, 49, 56, 58, 249, 251
Schnitzlein, K., 97
Schubertb, M., 108
Schwartz, J. G., 51, 191
Schweich, D., 92
Sederman, A. J., 34, 51, 53, 121, 128, 132, 140−141, 186
Sedricks, W., 87
Seguin, D., 27, 130
Shah, Y. T., 78, 93, 104, 191, 204
Sharma, S. K., 33, 204
Sherwood, T. K., 58, 84, 244−245
Shigarov, A. B., 176, 190
Shiiba, Y., 176
Shirai, T., 64
Sicardi, S., 33, 39, 51
Sie, S. T., 1, 29, 120, 180
Silveston, P. L., 78, 100, 192−193
Sims, B. W., 94
Smith, J. M., 26, 51−52, 58−60, 87−88, 90, 92−93, 174, 185, 187−188, 191, 204, 231
SØrensen, J. P., 127, 129
Spalding, D. B., 131

Specchia, V., 29, 32, 46, 55, 57, 59, 64, 66−67, 160−161
Spedding, P. L., 121
Spencer, R. M., 121
Sporka, K., 89, 176
Stanek, V., 176
Steven, M., 179
Stegeman, D., 191
Steiner, K., 95
Stephenson, J. L., 121
Stewart, W. E., 121, 127, 129
Stit, E. H., 142
Storck, A., 27, 57−58
Stradiotto, D., 193
Stuber, F., 94
Suekane, T., 128, 132−133
Sun, C. G., 146, 188
Sundaresan, S., 32, 146, 149, 159−160
Suwanprasop, S., 96
Sylvester, W. D., 185, 188
Szady, M. J., 149, 159−161
Szlemp, A., 33

T

Talmor, E., 29−30
Tan, C. S., 58−59, 61, 87
Taskin, M. E., 142
Taylor, R., 191, 237−245, 248−249
Thompson, K. E., 146
Thyrion, F. C., 55, 61
Tobis, J., 127−128
Tobolski, A., 193
Tomita, T., 64
Toppinen, S., 101
Tosun, G., 38, 61
Trambouze, P., 94
Trudell, C., 64
Tryggvason, G., 150
Tsochatzidis, N. A., 33
Tukac, V., 95

U

Ulbrecht, J., 57−59
Unverdi, S. O., 150

V

Valerius, G., 94, 98
Van, den, Bleek, C. M., 120, 128

Vander-Heyden, W. B., 146
Van, Rooijen, F. E., 191
Van, Swaaij, W. P. M., 60, 191
Varma, A., 64
Varma, Y. B. G., 32, 37–38, 42, 46, 160
Vasalos, I. A., 95
Vergel, C., 94
Villermaux, J., 60, 191

W

Wakao, W., 63
Wammes, W. J. A., 32, 34, 37–39, 43–44, 48–50, 55, 87
Wanchoo, R. K., 33, 63
Wanchoo, R. K., 33, 63
Wang, L., 175
Wang, Y., 35
Watson, P. C., 97, 177
Wauquier, J. P., 94
Weekman, V. W., 26, 63–65, 185
Weger, E., 51, 191
Wehner, J. F., 102
Weijer, S., 191
Wensrich, C. M., 121, 123, 125

Westerterp, K. R., 32, 34, 37–39, 43–44, 48–50, 55, 87, 191
Wild, G., 35, 44, 49, 55–56, 64, 66, 187
Wilhelm, A. M., 52, 174
Wilhelm, R. H., 102
Wolfstein, M., 131
Worley, Jr, F. L., 33, 57
Wu, Q., 96
Wu, Y., 52, 121
Wu, Y. X., 96, 175, 182

Y

Yamada, H., 108
Yin, F. H., 146, 188
Yokouchi, Y., 128
Yue, P., 126

Z

Zeiser, T., 128
Zhang, S., 175
Zhang, H., 126
Zhang, X., 175
Zhao, Z., 159, 239
Zhu, X., 98

Subject Index

A

Adiabatic reactor, 104–105
Axial dispersion, 25, 60–63, 68, 93–96, 99, 102, 191, 249

B

Bubble column, 9–11, 55, 242
Binary spherical system, 125
Biot number, 84
Body centered cubic, 129
Bondestine number, 61
Bubbly flow, 27, 29–31, 48, 67, 237–238, 248

C

Capillary forces, 13, 15, 37, 48, 86, 145, 148–148, 155, 161–162, 176, 192, 219, 253
Churn turbulent flow, 237–238
Complex reactions, 82, 92, 100, 105, 112, 196, 216, 223, 232
Computational fluid dynamics, 19, 68–69, 140, 171, 214, 223
Computer assisted tomography, 186
Conversion, 8, 10–11, 15, 30, 45, 49, 61, 63, 77, 86, 89, 91, 99–100, 102–103, 105, 108, 111, 171, 174–176, 194, 200, 202, 204, 211, 214–215, 219–222, 224, 229–233, 247–248
Corrugated Packing, 124
Cylindrical particle, 142, 180–181, 187

D

Diluted bed, 100
Dispersion coefficient, 60–63, 77, 93, 164, 190–191
Dynamic liquid holdup, 12–13, 33–34, 45–47, 55, 66–67

E

Ejector loop reactor, 9–11
Esterification, 3
Effectiveness factor, 80, 82–89, 98–99, 102, 106, 141, 163, 174, 180, 212, 218, 227, 233
Ergun's constant, 144, 160
Eulerian-Eulerian model, 18, 146, 165
Exothermic reactions, 10–11, 13, 87, 89–91, 108, 199, 214, 222–223, 231, 236
External mass transfer, 21, 84, 86–87, 100, 182, 201, 212, 242
Extrudates, 4, 7, 57, 61, 121, 181

F

Face centered cubic (FCC), 127, 129, 136, 138, 140–141
Fibers, 125
Flow modeling, 5, 16, 117–165
Flow regime, 12–13, 18, 21, 25–69, 77, 79, 109, 117–118, 127, 130, 134, 136, 140, 147, 154, 160, 162, 171–172, 186, 192, 205, 212–213, 237–238, 245, 248–250
    transition, 26, 32–34, 37–39, 68
Fluid-fluid interaction, 147, 192
Froude number, 53F-T synthesis, 3

G

Gas limiting reactions, 21, 189, 193–194
Gas-liquid interfacial area, 151
Gas-liquid mass transfer, 218, 240
Gas-solid mass transfer, 78
Gas to liquid reactor, 103, 163
Gaussian distribution, 122–123
Gauze packing, 124
Grid type packing, 124

H

Heat transfer, 4, 7, 9−11, 15, 18, 25−26, 40,
    63−68, 77, 84, 86, 89, 99, 103−105,
    107, 109, 118, 121, 127, 131−132,
    141−142, 164, 171, 173, 196−198,
    201, 211, 225−226, 235, 247,
    251−252
Honeycomb, 224, 236−237
Hydrodynamics, 16, 21, 25−69, 77−80,
    112, 117−118, 157−158, 164,
    171−172, 174, 178, 180, 182, 187,
    192, 196, 198, 206, 213, 218, 223,
    225−226, 228, 233, 236−237, 248,
    251−252
Hydrogenation, 1−3, 20, 22, 55, 63, 87,
    89−102, 105, 108, 110, 175−179,
    189−191, 194, 200−201, 212−224,
    246, 251
    reaction, 2, 91, 102, 190, 201, 212−214,
    246, 251
Hydroprocessing, 20, 22, 77, 90, 95, 102,
    164, 184, 214, 224−231
Hysteresis, 41, 46, 89−91, 154, 173,
    177−179, 192, 206, 214, 223, 253

I

Interstitial velocity, 35
Intraparticle mass transfer, 80−82, 98−99,
    102, 105−106, 112, 216, 218, 226,
    231

L

Laboratory scale reactor, 195−198, 200,
    228−230, 233
Langmuir Hinshelwood model, 78, 82,
    97−100, 246
Liquid distribution, 9, 11, 13, 15, 52, 60, 79,
    86, 89, 91, 145, 148, 155, 172−173,
    179, 182, 184−185, 187−190,
    192−193, 195, 199, 201, 235−237,
    248, 253
Liquid distributor, 7, 52, 182−184
Liquid holdup, 11−13, 21, 25−26, 31, 33−34,
    36−38, 45−50, 52−53, 55, 66−68,
    88−89, 93, 100−101, 118−119, 145,
    148, 151−157, 162, 164, 172, 174,
    180, 191−192, 200−201, 205, 228,
    230, 242, 249, 251
Liquid limiting reactions, 21, 104, 193−194

Liquid mal-distribution, 196
Liquid-solid mass transfer, 81, 218, 243−244

M

Monolith reactor, 5, 9, 212, 235−247
Micro reactor, 1, 8, 197, 200−201, 247−251
Multi-scale modeling, 16−17, 21, 25, 39, 252
Magnetic resonance imaging (MRI), 5, 34,
    51, 55, 91, 124, 128, 178−179
Meso-scale model, 19, 119, 145, 149−154
Micro reactor, 1, 8, 197, 200−201, 247−251
Micro-scale model, 17
Momentum exchange terms, 36−37,
    146−147
Monolith reactor, 5, 9, 212, 235−247
Mono-size spheres, 125
Multiplicity, 15, 79, 89, 176, 205−206, 214,
    224

N

Navier−Stokes, 130
Non-isothermal reactor, 99
Non-prewetted, 40−42, 155−157, 186, 192
Nusselt number, 109, 136, 141, 245, 262

O

Oxidation reaction, 2, 77, 93, 192, 201,
    231−235, 246
Overall effectiveness factor, 80, 82−85,
    87−88, 180
Overall reaction rate, 173−175
Oxidation, 1−2, 20, 22, 55, 63, 77, 93−96,
    100−101, 108, 192, 200−201, 211,
    231−233, 246, 251

P

Packed bubble column reactor, 9−11
Petroleum processing, 200
Partially wetted catalyst, 79, 86−89, 218
Particle Reynolds number, 27, 127, 129−131,
    133−134, 136−138, 140, 262
Particle-fluid heat transfer, 41
Peclet number, 61, 66, 93
Periodic flow operation, 193, 199
Piece wise linear interface calculation,
    150
Pilot scale reactor, 16, 197, 201

Polylobes, 144, 180−181
Porosity, 4−5, 13, 18−19, 34, 36−37,
40−42, 52, 60−61, 66, 97, 118,
121−125, 127−129, 132, 141,
144−146, 148−149, 157,
163−165, 180−182, 192, 196, 198,
226, 228
Porosity variation, 18, 122, 163−164, 182,
196, 198, 228
Pressure drop, 40−45, 119, 198−199,
240−243
Prewetted, 40−42, 149, 154−157, 186−187,
192
Pseudo-homogeneous reactions, 93,
102−104, 214, 251
Pulse flow, 27−35, 37−39, 41−43, 46−48,
64, 67, 154, 160, 162, 192, 200,
249−250

Q

Quadrilobes, 4, 180−181

R

Reactor engineering, 5, 12−22, 25, 117−118,
146, 165, 173, 195, 203, 205−206,
212, 230, 237, 244−246, 250,
252−253
Random packing, 8, 121, 123−124, 127
Rasching rings, 181
Reaction engineering, 5, 16, 19, 21, 25, 69,
77−112, 117, 171, 190, 195,
203−206, 212−215, 223, 225, 228,
232, 239, 248, 250−252
Reactor performance, 5, 7, 13, 16, 20−21, 50,
78−79, 80, 87, 90−102, 104, 112,
117, 119, 164−165, 171−207
Residence time distribution, 12, 22, 60,
118−119, 164, 172−173, 190−192,
247
Reynolds number, 27, 30−31, 42, 48,
52, 56, 58−59, 61−62,
66−67, 127−134, 136−138,
140, 238
Rhombohydral, 129, 136−141

S

Stirred tank, 55, 97, 242
Slurry reactor, 9−11, 98, 197, 214

Saturated liquid holdup, 34, 45, 48−49, 148,
158−161
Scale-down, 5, 22, 118, 195−197, 206,
207
Scale-up, 1, 5, 12, 22, 25, 63, 68, 79, 89, 112,
118, 124, 164−165, 171−207,
212−213, 225, 230, 232−233, 235,
247, 251−252
Schmidt number, 56, 58
Selectivity, 11, 63, 86, 89, 100, 103,
204−205, 214, 220
Sherwood number, 58, 84, 244
Simple line interface calculation, 150
Single phase flow, 21, 26−27, 46, 62, 97, 118,
126−128, 130, 132, 149, 165, 241,
244
Slug flow, 237−238, 241, 249
Specific surface area, 46, 48, 119, 180
Spray flow, 13, 27, 29, 31−32, 160, 162
Static liquid holdup, 12, 36, 45, 205
Structured packing, 5, 8−9, 15, 120−121,
124−126

T

Trickle bed reactor
applications, 212−235
configuration, 5
operation, 5, 11, 112
co-current, 78
counter current, 6
jacketed, 6
internally cooled, 6
Taylor flow, 237−242, 244−245,
248−249
Thiele modulus, 82, 84, 219
Three layered packing, 124
Tortuosity, 180
Trickle flow, 13, 19, 27−32, 37−38, 41−42,
46−48, 50, 55, 57−58, 62, 64,
66−67, 77−78, 100, 119, 154, 160,
162, 186, 200, 249
Turbulence model, 130, 152
Turbulent kinematic viscosity, 130
Turbulent kinetic energy, 258
Turbulent Prandtl number, 134
Turbulent viscosity, 130
Two phase flow, 23, 27, 43, 237, 241−242

V

Volume of fluid, 18−19, 165

Vorticity, 151
Weber number, 42, 49

W

Wetting, 4, 9, 11−13, 15, 19, 21−22, 26,
    30−31, 45−46, 49−55, 57−58, 60,
    62, 64, 67, 74, 79, 86−90, 93−105,
    145, 149, 163, 172−178, 180, 182,
    185−186, 189, 190−193, 195−196,
    199, 201−202, 205, 214−215,
    219−220, 222−223, 226−227, 233,
    235, 251, 253
Wetting efficiency, 21, 40, 45, 49, 51−54, 77,
    86, 97−98, 101, 185, 201, 219−220,
    227

DATE DUE

~~INTERLIBRARY LOAN~~			New
			12/16/14

Demco, Inc. 38-293